EVALUATING MILITARY ADVERTISING AND RECRUITING

Theory and Methodology

Committee on the Youth Population and Military Recruitment—Phase II
Paul R. Sackett and Anne S. Mavor, *Editors*

Board on Behavioral, Cognitive, and Sensory Sciences

Division of Behavioral and Social Sciences and Education

NATIONAL RESEARCH COUNCIL
OF THE NATIONAL ACADEMIES

THE NATIONAL ACADEMIES PRESS
Washington, D.C.
www.nap.edu

THE NATIONAL ACADEMIES PRESS 500 Fifth Street, N.W. Washington, D.C. 20001

NOTICE: The project that is the subject of this report was approved by the Governing Board of the National Research Council, whose members are drawn from the councils of the National Academy of Sciences, the National Academy of Engineering, and the Institute of Medicine. The members of the committee responsible for the report were chosen for their special competences and with regard for appropriate balance.

The study was supported by Contract No. M67004-00-C-0030 between the National Academy of Sciences and the U.S. Marine Corps. Any opinions, findings, conclusions, or recommendations expressed in this publication are those of the author(s) and do not necessarily reflect the view of the organizations or agencies that provided support for this project.

Library of Congress Cataloging-in-Publication Data

National Research Council (U.S.). Committee on the Youth Population and Military Recruitment.

Evaluating military advertising and recruiting : theory and methodology / Committee on the Youth Population and Military Recruitment--phase II ; Paul R. Sackett and Anne S. Mavor, editors.

p. cm.

Includes bibliographical references and index.

ISBN 0-309-09127-6 (hardcover) -- ISBN 0-309-52947-6 (pdf)

1. United States--Armed Forces--Recruiting, enlistment, etc. 2. Manpower--United States. 3. Youth--United States. I. Sackett, Paul R. II. Mavor, Anne S. III. Title.

~~UB323.A5 2004~~

355.2'2362'0973--dc22

2003027271

Additional copies of this report are available from The National Academies Press, 500 Fifth Street, N.W., Lockbox 285, Washington, D.C. 20055; (800) 624-6242 or (202) 334-3313 (in the Washington metropolitan area); Internet, http://www.nap.edu.

Printed in the United States of America.

Suggested citation: National Research Council. (2004). *Evaluating Military Advertising and Recruiting: Theory and Methodology.* Committee on the Youth Population and Military Recruitment—Phase II. Paul R. Sackett and Anne S. Mavor, editors. Board on Behavioral, Cognitive, and Sensory Sciences, Division of Behavioral and Social Sciences and Education. Washington, DC: The National Academies Press.

COMMITTEE ON THE YOUTH POPULATION AND MILITARY RECRUITMENT—PHASE II

PAUL R. SACKETT (*Chair*), Department of Psychology, University of Minnesota, Twin Cities
DAVID J. ARMOR, School of Public Policy, George Mason University
JERALD G. BACHMAN, Institute for Social Research, University of Michigan, Ann Arbor
JOHN EIGHMEY, Greenlee School of Journalism, Iowa State University, Ames
MARTIN FISHBEIN, Annenberg School for Communication, University of Pennsylvania
CAROLYN SUE HOFSTRAND, Taylor High School, Volusia County, Florida
PAUL F. HOGAN, The Lewin Group, Fairfax, Virginia
JAMES JACCARD, Department of Psychology, State University of New York, Albany
CAROLYN MADDY-BERNSTEIN, Education Consultant, Oro Valley, Arizona
CAROL A. MUTTER, LTG, retired, U.S. Marine Corps
LUTHER B. OTTO, emeritus, Department of Sociology, North Carolina State University, Raleigh
WILLIAM J. STRICKLAND, HumRRO, Alexandria, Virginia
NANCY T. TIPPINS, Personnel Research Associates, Dallas, Texas
JOHN T. WARNER, Department of Economics, Clemson University

BRUCE R. ORVIS, *Committee Consultant*, RAND Corporation, Santa Monica, California

ANNE S. MAVOR, *Study Director*
MARILYN DABADY, *Senior Research Associate*
WENDY E. KEENAN, *Senior Project Assistant*

Preface

The Committee on the Youth Population and Military Recruitment was established by the National Research Council (NRC) in 1999 in response to a request from the U.S. Department of Defense. The impetus for the study was the recruiting problems encountered by the Services in the late 1990s. The central question is how to attract qualified youth to serve their country and, if necessary, be willing to put themselves in harm's way. Although military missions have diversified since the end of the cold war, the primary function of the Services remains the provision of the nation's warriors and protectors.

The charge to the committee was to provide information about the demographic characteristics, skill levels, attitudes, and values of the youth population; to examine options available to youth following high school graduation; and to recommend various recruiting and advertising strategies and incentive programs to encourage enlistment. In the first phase of its work, the committee confirmed that propensity for military service was declining. In 2002, the committee published *Attitudes, Aptitudes, and Aspirations of American Youth: Implications for Military Recruitment.*

One outcome of the first phase was the recognition that current military research on advertising and recruiting often lacked long-term objectives and coordination across relevant research topics and methodologies. As a result, the committee embarked on a second phase: to develop an evaluation framework to assist the Defense Department and the Services in making informed decisions on the effectiveness of various recruiting policies and mixes of recruiting resources. This report is the product of the committee's second phase of study.

Several individuals provided the committee with useful information on evaluation strategies and current Defense Department advertising programs. First, we would like to thank Robert Hornik, Annenberg School of Communication, University of Pennsylvania, and James Dertouzos, RAND Corporation, for their excellent presentations on research methodology. We would also like to extend our gratitude to Heather LeFevre and Jan Ross, of Mullen, for their presentation on the military's joint advertising program; to Jay Cronin, of J. Walter Thompson, for his presentation on Marine Corps advertising; and to Col. Greg Parlier, U.S. Army Aviation and Missile System Support, for the information he provided on the Army's recruiting and advertising programs.

We express appreciation to our sponsor, the Office of Assistant Secretary of Defense for Force Management Policy, for its interest and guidance. Particular thanks are due to Curt Gilroy and to Jane Arabian.

In the course of preparing this report, each member of the committee took an active role in drafting chapters, leading discussions, and reading and commenting on successive drafts. We are deeply indebted to all for their broad scholarship and their cooperation and spirit.

The committee is also particularly indebted to Bruce Orvis, RAND Corporation, who served as a consultant to the committee. He drafted the chapter on determining optimal types of incentives and provided many insights regarding material in other chapters of the report.

This report has been reviewed in draft form by individuals chosen for their diverse perspectives and technical expertise, in accordance with procedures approved by the NRC's Report Review Committee. The purpose of this independent review is to provide candid and critical comments that will assist the institution in making its published report as sound as possible and to ensure that the report meets institutional standards for objectivity, evidence, and responsiveness to the study charge. The review comments and draft manuscript remain confidential to protect the integrity of the deliberative process. We wish to thank the following individuals for their review of this report: Morton G. Ender, Department of Behavioral Sciences and Leadership, United States Military Academy; Lawrence Goldberg, Cost Analysis and Research Division, Institute for Defense Analyses; Stanley A. Horowitz, Cost Analysis and Research Division, Institute for Defense Analyses; James Hosek, Economics and Statistics Group, RAND Corporation; W.S. Sellman, Human Resources Research Organization; and Bruce G. Vanden Bergh, Department of Advertising, Michigan State University.

Although the reviewers listed above have provided many constructive comments and suggestions, they were not asked to endorse the conclusions or recommendations nor did they see the final draft of the report before its release. The review of this report was overseen by Robert Linn,

School of Education, University of Colorado. Appointed by the National Research Council, he was responsible for making certain that an independent examination of this report was carried out in accordance with institutional procedures and that all review comments were carefully considered. Responsibility for the final content of this report rests entirely with the authoring committee and the institution.

Staff of the National Research Council made important contributions to our work in many ways. We extend particular thanks to Marilyn Dabady for her outstanding efforts as a senior research associate. We are also grateful to Wendy Keenan, the committee's senior project assistant, who was indispensable in organizing meetings, arranging travel, compiling agenda materials, and in managing the preparation of this report.

Paul R. Sackett, *Chair*
Anne S. Mavor, *Study Director*

Contents

Executive Summary

In the late 1990s, the U.S. armed forces struggled to meet their recruiting goals and in some cases fell short. This led to the question of how the recruit-ing process and the recruiters' job could be better supported in order to ensure that force strength, force quality, and the required skill mix of personnel will be available to meet ever-changing security and defense challenges. Military officials recognized that a fundamental understanding of the youth population and of the effectiveness of various advertising and recruiting strategies used to attract them would be extremely valuable in addressing these questions.

As a result, in 1999, the Department of Defense asked the National Academy of Sciences, through its National Research Council, to establish the Committee on the Youth Population and Military Recruitment. The committee is composed of 14 experts in the areas of military manpower, military sociology, psychology, adolescent development, survey methodology, behavioral theory, economics, and advertising and communication.

During the first phase of its study, the committee examined long-term trends in the youth population and evaluated policy options that could improve the propensity for and enlistment in the Services. In our report, *Attitudes, Aptitudes, and Aspirations of American Youth*, the committee confirmed the decline in propensity for military service among youth and identified several correlates, especially the trend in increasing college enrollments.

The committee observed that current military research on advertising and recruiting often lacked long-term objectives and coordination across relevant research topics and methodologies. In this second phase, the committee has developed an evaluation framework to assist the Department of Defense and the Services in making informed decisions on the effectiveness of various recruiting policies and mixes of recruiting resources. This report is the product of the committee's second phase of study.

THE APPROACH

The committee has identified several areas requiring more intensive study that might benefit from drawing on a variety of methodological approaches:

- Monitoring trends in youth attitudes, values, and propensity using surveys (Chapter 3);
- Planning advertising using generative and experimental approaches (Chapter 4);
- Determining optimal levels of advertising and recruiting resources and assessing the timing and levels of joint and Service-specific Advertising based on data from past or current programs using econometric methods (Chapters 5 and 6);
- Determining optimum types of incentives using a combination of focus groups, surveys, and experimental approaches (Chapter 7); and
- Performance management of recruiters again using a combination of methodologies (Chapter 8).

Rather than focusing on the strengths and weakness of the various methods with the intent of identifying some methods as per se superior to others, the committee's proposed evaluation framework is based on the fundamental notion that different research designs and the associated methodologies are suited to address different types of research questions. Table ES-1 provides examples of common research questions that emerge in the context of military advertising and recruiting.

The framework has two dimensions. The first dimension differentiates between an existing or new incentive, program, or activity; the second dimension differentiates between three types of assessments. One type of assessment occurs in situations in which the outcome is not specified and the audience is asked to indicate the attractive and unattractive features; the second type deals with attitudes or behavioral intentions toward military enlistment; and the third deals with actual behavior, such as signing a contract with a recruiter. The cells in the framework list example types of questions and identify the method most appropriate for addressing each.

TABLE ES-1 Evaluation Framework

	Outcome Not Specified A Priori	Specific Attitudes or Behavioral Intentions	Actual Behavior
New programs	Question: "What does a target audience see as attractive or unattractive features of a program?" Method: focus groups; unstructured or open-ended surveys and interviews (Chapters 4 and 7)	Question: "What is the effect of a program on specified attitudes or behavioral intentions?" Method: surveys; experiments; quasi-experiments (Chapters 3, 4, 7, 8)	Question: "What is the effect of a proposed new program on enlistment?" Method: experiments; quasi-experiments (Chapters 7 and 8)
Existing programs	Same as above	Same as above	Question: "What is the effect of an existing program on enlistment?" Method: econometric modeling (Chapters 5 and 6)

CONCLUSIONS AND RECOMMENDATIONS

Monitoring Trends in Youth Attitudes, Values, and Propensity

In its first report, the committee concluded that propensity to enlist is a major direct determinant of actual enlistment. Thus, increasing propensity should be an important goal of the military. Monitoring surveys are well suited to measuring trends in propensity and examining the factors that contribute to changes over time. Implementation of useful monitoring surveys requires multiyear funding commitments and large samples of respondents. In constructing a monitoring survey to track propensity, the questionnaire must include a complete set of salient beliefs about the positive and negative consequences of joining the military. Other important content areas are the values attached to various outcomes, the expectation that a particular outcome is more likely to be achieved in a military versus a civilian job, and the barriers or facilitators for enlistment decisions. In order to ensure complete coverage of these attitudes and beliefs, good survey designs should also conduct certain types of preliminary or exploratory studies. The committee recommends that survey research examining propensity be designed to incorporate the key determinants of propensity and that it be designed to permit analysis at the individual level.

The committee proposes a program of survey research involving a commitment of at least five years. The committee recommends that consideration be given to undertaking a school-based survey, using cohort-sequential design, in which students are sampled in the 11th and possibly the 12th grade and regularly resurveyed until the age of 23 or 24.

Planning for Advertising

The purpose of advertising is to distribute information designed to influence consumer activity in the marketplace. In the military the purpose of advertising is twofold: to increase propensity to chose military service and to increase the likelihood of an individual choosing to join one Service over another. In the past, much of the research on military advertising has focused on evaluating the amount and cost of advertising, rather than on evaluating the effects of advertising content on targeted beliefs and values. In its first book, the committee found that intrinsic factors, such as duty to country, should be given increased weight in military advertising.

In a strategy for developing and evaluating a range of message strategies, the first step is to track the competitive environment for military recruitment to detect factors affecting youth understanding and views of military service. The second step is to examine the beliefs, goals, and language of audience members using focus groups, interviews, and surveys. The third step is to develop and test a range of message strategies using experimental and quasi-experimental research designs. The fourth step is to allocate resources to various message strategies. Decision making during this step is informed by experiments and in-market testing.

The committee recommends a program of research that follows these steps. Specifically, a program of research should begin with generative techniques to understand the concepts and language used by youth in considering alternative courses of action (e.g., higher education versus military service) and continue with survey research that measures the full range of beliefs, attitudes, and values that emerge as linked to these alternative courses of action. The committee also recommends that advertising message strategies be evaluated in terms of their effects on targeted beliefs and values. Such evaluation should make use of experimental designs in controlled setting and small-scale, in-market experiments. Advertising should be evaluated in terms of thematic content in order to determine whether its effects vary by content as well as by impressions and expenditures.

Determining Optimal Levels of Advertising and Recruiting Resources

Many of the most important determinants of enlistment supply, as well as the cost and effectiveness of existing recruiting resources and the trade-offs among them, have been addressed by a well-developed body of econometric research. Econometric methods are most appropriate for studying existing programs and for developing estimates based on natural variation in resources and outcomes.

There have been several studies over the past 20 years on the effects of recruiting and advertising on enlistment. The variability among these studies suggests that a consistent methodology has not been incorporated. Based on our review of the literature, the committee concludes that:

- Recruiter productivity varies with experience, and hence sudden changes in the size of the recruiting force result in declines in average experience. Failure to incorporate recruiter experience into models of recruiter effects may bias study results.
- Recruiter incentives have been incorporated in supply models via recruiters' quotas, based on the assumption that increasing recruiting quotas increases effort. A more complete and realistic model of recruiter incentives is needed.
- Research to date has not incorporated the effects of Reserve competition on active-duty recruiting.

The committee also concludes that the functional forms (i.e., the shape of the relationship between the recruiting incentive and enlistment) of econometric supply models have been relatively restricted. The underlying assumptions (e.g., that each additional advertising dollar has the same effect regardless of the level of total expenditure) may not be correct, and an examination of more flexible functional forms would be fruitful. The committee therefore recommends that research on supply models make use of flexible functional forms, rather than imposed functional forms.

A theme underlying all of the suggested areas for improvement is the need for better data, consistently collected and retained over time. Ideally, these data should include enlistment contract and accession data, by level of qualification and Service, at the lowest reasonable level of aggregation and time period. The data should also include information on the resources and incentives that have been applied and the external factors that were in effect during the period, as well as indications of recruiting policies, incentives, and quotas.

Timing and Levels of Joint and Service Advertising

Econometric methods are also applicable to the questions of (1) whether there is a minimum level of advertising necessary for a cost-effective recruiting program even if it is not necessary to achieve current enlistment contract goals and (2) what are the proper levels of joint and Service-specific advertising. Quasi-experimental methods can also be employed to address the second question.

We describe the conditions under which it would be cost-effective to advertise in the interests of future enlistment supply and review research to date on this question. While this research suggests that advertising may have effects only for a short period of time, the data available to previous researchers are limited as they do not permit the examination of time-lagged and nonlinear effects within a time period. In the committee's view, it would be a serious mistake to view the available research as sufficient grounds for drawing conclusions one way or another about the effects of current advertising on future outcomes. The committee recommends a focused effort to maintain advertising data in a systematic way for purposes of estimating a supply curve that incorporates the potential for both time-lagged and nonlinear advertising effects. The committee also recommends a program of research, incorporating quasi-experimental methods, to examine advertising effects over an extended period of time.

Regarding the question on the appropriate levels of joint and Service-specific advertising, in our judgment, certain types of advertising themes, such as generic themes designed to increase overall propensity, are best done as a joint program, while advertising themes featuring specific benefits of military service are best done in the Service program. What we do not know is what level of advertising funding should be allocated to joint programs. We therefore recommend a program of research aimed at examining the effects and cost-effectiveness of information-oriented versus values-oriented advertising in joint and Service advertising programs.

Determining Optimal Types of Incentives

The Services have many types of enlistment incentive programs in operation today, most of which are aimed at highly qualified recruits. Current incentives include education benefits as well as enlistment bonuses for various types of jobs. The Services are experimenting with other incentives, particularly those that forge a closer connection between military and college pursuits, and also those that offer differing lengths of active-duty and Reserve tours.

We review research on various incentive programs, both old and new, in terms of the different types of research designs that have been or are being used for studying this topic: survey, experimental, and econometric. The central message is that each of the evaluation methodologies discussed in previous chapters can play a useful role in addressing different questions that policy makers may ask about current or proposed incentives. It is also important to note the value of combining approaches in examining a particular program. For example, focus groups can be used to explore prospective enlistment options, which are then tested with a large survey of youth or in pilot tests employing experimental designs.

Performance Management of Recruiters

Recruiter performance management encompasses the range of issues and decisions that face Service recruiting managers as they organize to meet their mission. Service recruiting managers establish systems to select and train recruiters, to open recruiting offices in specific locations, to establish production goals, to motivate recruiters with various incentive programs, and to monitor and assess recruiter performance. In some of these areas, such as developing recruiter selection programs, there is a large body of ongoing research in both the military and civilian sectors. In other areas, such a developing effective incentive programs, research efforts are minimal.

In many respects, the problems of performance management faced by the military are no different from the problems faced by private industry. However, the environments are distinctly different, and the military faces many restrictions that do not apply in the civilian sector. As a result, some of the existing research from the professional literature will be of limited use. Ideally, the military should undertake a continuous and systematic evaluation of each aspect of performance management system individually and in combination. Specifically, the committee recommends:

- Continued research on the development of effective recruiter selection strategies, in conjunction with a consideration of career incentives for service as a recruiter.
- Expansion of the Services' evaluation of overall training of recruiters to include the study of other informal development opportunities. In particular, assessment and improvement of the supervisory and coaching

skills (to include on-the-job training) of those who train recruiters may be a fruitful approach.

- A program of research aimed at evaluating the effects of goals on recruiter behavior and outcomes.
- Research to develop a complete model of recruiter performance and to develop performance appraisal instruments and feedback processes based on this model.

1

Introduction

During the late 1990s, the armed forces began having difficulty meeting recruiting targets, particularly for highly qualified recruits—youth with high school diplomas and above-average aptitudes who are essential for an effective fighting force. Recruiting problems were most severe during 1998 and 1999, when most of the Services experienced accession shortfalls. These shortfalls were especially alarming to military planners, given the major force reductions of the 1990s and the lower accession requirements that followed (Sellman, 1999).

In 1999, at the request of the U.S. Department of Defense (DoD), the National Research Council formed the Committee on the Youth Population and Military Recruitment. During Phase I of its study, the committee was asked to examine long-term trends in the youth population and evaluate policy options that could improve the propensity for and enlistment in the Services. In its Phase I report, *Attitudes, Aptitudes, and Aspirations of American Youth*, the committee confirmed the decline in propensity for military service among youth and identified several correlates, especially the increasing trend in college enrollments. The report made specific recommendations about changing advertising programs and recruiting policies to improve propensity and enlistment.

During the earlier phase, the committee observed that military research on advertising and recruiting programs is often opportunistic and fragmented, lacking coordination and long-term objectives. In this report, the committee aims to fill this gap by proposing a comprehensive evaluation framework to assist and improve research on recruiting policies and advertising programs. This report contains results, conclusions, and recommendations from the committee's Phase II study.

PURPOSE OF THE STUDY

The primary objective of this study is to help DoD improve its research on advertising and recruiting policies. It is anticipated that in the coming decade DoD will field and test new advertising and recruiting initiatives designed to improve the recruiting outlook and help avoid the shortfalls of the past decade. In order to discover the most promising policies, in the committee's view the department needs a comprehensive research and evaluation strategy based on sound research principles that will ensure valid, reliable, and relevant results.

Good research should begin with developing a clear statement of the problems to be studied, because the nature of the policy questions influences and shapes the most appropriate research designs. While a given research question can be studied using a variety of approaches and techniques, some research designs in our view are more promising than others for particular research questions. For example, focus groups are most useful when determining what a target audience finds attractive about a program, whereas econometric methods are appropriate when studying the effects of an existing program on enlistments.

Good research requires more than sound methodology; it must also be grounded in solid theory and valid concepts. In the case of recruiting and advertising policies, the most important theory concerns individual decision-making processes, particularly the factors than influence job and career decisions. The research program proposed here reviews and develops the most relevant theories for enlistment decisions.

Finally, advertising and recruiting research must produce results that are useful to military planners. Research results must be available in a timely manner to help planners make decisions among competing proposals and to assist during the programming and budgeting cycles of the department. In the committee's view, the research and evaluation studies proposed in the following chapters can meet these needs.

The remainder of this introduction is an overview of issues and methods and proposes a framework for selecting the most appropriate method to address various types of research questions.

THE EVALUATION PROBLEMS

The committee's work during Phase I led us to conclude that there are a number of critical problem areas or topics needing more intensive study. Some problems arise because of the need for ongoing, up-to-date information that can serve as early-warning indications of potential recruiting problems or that can point to areas in which improvements are needed. Other problem areas are important because they are central to improving the overall recruiting climate. We have selected six areas to be the central focus of this report.

Monitoring Trends in Youth Attitudes, Values, and Propensity

The issue of youth attitudes has, in fact, received considerable emphasis from DoD. Youth attitudes, values, and propensity for military service were monitored by the Youth Attitude Tracking Studies (YATS) from 1975 to 1999. A new effort to survey youth attitudes on a quarterly basis began in 2001 as part of the Defense Marketing Research program (Sellman, 2001).

Further monitoring of youth attitudes is important for several reasons. First, this information can be used as leading indicators of future changes in enlistment behaviors. For example, YATS documented a downturn in youth propensity that presaged the 1998-1999 recruiting shortfalls by at least six years.

Second, the YATS methodology was not optimal, as was documented in the committee's letter report of June 2000. The once-per-year format hampers the assessment of effects on propensity arising from specific events, such as the Persian Gulf and the latest Iraq wars. In addition, YATS administered different attitude and value questions to different subsamples of youth at different times, making it very difficult to analyze the relationship between attitudes or values and enlistment propensity. Finally, it is not clear that YATS adequately covered the domain of values and attitudes that influence career decisions.

DoD should continue to place a high priority on studies of short- and long-term trends in youth attitudes, values, and propensity. The new Defense Marketing Research program (Office of Assistant Secretary of Defense [Force Management Policy], 2000) has improved the frequency of administration, but issues remain regarding the collection of trend data on key values and attitudes relevant to military service. The methodology used in these studies should ensure timely, reliable, and relevant information that can be used for anticipating potential problems in enlistment supply.

Developing Effective Advertising Themes to Increase Youth Propensity

The committee's Phase I report concluded that current advertising strategies and themes are not designed to maintain a base level of propensity, and this may be contributing to the decline in youth propensity for military service in light of alternatives (e.g., college, employment).

Accordingly, one of our most important recommendations was to develop effective advertising themes to help increase the propensity of youth for military service.

Solving this problem requires two types of research. The first is development of promising advertising themes using generative research techniques; the second is thoroughly and rigorously testing the market impact of the selected themes. Again, the Services and DoD have conducted research in this area, but many improvements can be made in the methodologies applied to date, and these improvements can have a real impact on the development of effective advertising themes and message delivery strategies.

Optimal Levels of Recruiting Programs and Resources

The Services have many existing policies and programs designed to maintain adequate levels of enlistment supply. These include traditional and on-line advertising, various incentive packages (education benefits and enlistment bonuses), and number and location of recruiters. The policy problem is how to allocate these resources to provide effective and efficient approaches for maintaining a given supply of highly qualified recruits.

Results of optimal mix studies are extremely useful for allocating resources among various existing programs during the budget planning cycles. While the Services have conducted some research in this area, many existing studies have not adopted the strongest research designs or incorporated the most reliable data. Improving the quality and reliability of this research would have significant payoff in helping to provide the most efficient mix and level of existing recruiting resources.

Optimal Investment in Joint and Service Advertising

The timing and level of advertising are somewhat related to both advertising planning and optimizing the levels of recruit programs and resources, but we discuss them separately to emphasize two important areas of research that have not been adequately studied to date.

First is the issue of level—the minimum level of advertising needed to maintain youth propensity even during periods when enlistment supply is good. Service enlistment goals can fluctuate from year to year for various policy reasons, and enlistment supply is also influenced by changes in external factors (e.g., unemployment). During good supply years, military planners often reduce recruit advertising below levels necessary for maintaining the propensity base, thereby creating an advertising vacuum that lowers propensity and supply in difficult years.

The second issue is the question of the optimal combination of joint (DoD-wide) and Service-specific advertising. Some advertising themes, particularly those designed to maintain or increase the base level of propensity (e.g., intrinsic values such as duty to country), may be accomplished more efficiently as a joint program, while other advertising themes (e.g., specific benefits of serving in a particular Service) are more appropriate for Service-specific programs. What we do not know, because of a paucity of research, is what combination of joint and Service advertising would be optimal for meeting both types of goals.

Improving Enlistment Incentives

The Services offer a wide assortment of enlistment incentives at the present time, including benefits for college education and enlistment bonuses for particular military jobs. For the most part, these incentives are similar to those introduced at the inception of the All Volunteer Force in 1974. Given the many competing alternatives facing youth today, particularly the increasingly popular college option, there may be other types of incentives that would enhance enlistment supply. For example, some Services are experimenting with programs that tie military duty more closely to college education, and other incentives being considered involve changing the length of active-duty and reserve commitments with the potential for enrollment in civilian national service. Research on new types of incentives, or new combinations of incentives, is very important to meet the increasing competition from college and other work opportunities.

Performance Management of Recruiters

Existing research has been very suggestive, but not definitive, about the impact of recruiter organization and performance on enlistment supply. By recruiter organization we mean such policies and practices as the number of recruiters, their geographic location, how they are selected from and returned to other military units, how they are trained, their pay and benefits, and the setting of recruiting goals or targets. While there is some existing research on recruiter organization and performance, comprehensive studies do not exist. There is potential for substantial payoffs from research that focuses on how the performance management of recruiters can increase enlistment supply in a very cost-effective manner.

THE EVALUATION FRAMEWORK

Our evaluation framework is based on the fundamental notion that different research designs are suited to answer different types of research questions. What appear to be disagreements about the relative merits of econometric models versus experiments versus survey methods tend to diminish if the question is framed as "What research approaches are best suited to answering the following research question?" rather than "What is the best approach to doing research?"

We outline a number of common types of research questions that emerge in the context of military recruiting. Note that this is not intended as a general or complete taxonomy of research questions; other types of questions may arise in other research contexts.

The research questions are placed in a two-dimensional framework (Table 1-1). The first dimension indicates whether the research question focuses on an existing incentive, program, or activity or a proposed new incentive, program, or activity (hereafter summarized generally as "program"). For example, an existing program might be a current advertising campaign, while a new program might be a proposed new enlistment incentive.

The second dimension differentiates among three types of assessments of new or existing programs. The first type focuses on settings in which the outcome variable of interest is not specified a priori. It addresses what a target audience sees as attractive and unattractive features of a current or proposed program. The second type of assessment focuses on specific attitudinal or behavioral intent variables, such as one's perceived benefits of military service or one's intention to enlist. It addresses the effect of existing or proposed programs on attitudes or behavioral intentions. The third type of assessment focuses on actual behavioral outcomes, such as an enlistment decision or contact with a recruiter. It addresses the effect of existing or proposed programs on a behavior of interest.

Each of the cells in this two-dimensional framework lends itself to some research methods more than others. Some cells may preclude the use of certain methods.

The framework identifies focus groups or open-ended surveys and interviews as the most viable method when the research question seeks opinions about and reactions to a current or proposed program. The use of such methods makes sense when the researcher has not specified in advance the attitudinal or behavioral variables of interest. The reactions obtained from individuals in focus groups may be completely unanticipated by the researchers. Such a strategy is perhaps most generally useful in the early stages of considering a new program, although it may be usefully applied to existing programs, particularly those that were imple-

TABLE 1-1 Evaluation Framework

	Outcome Not Specified A Priori	Specific Attitudes or Behavioral Intentions	Actual Behavior
New programs	Question: "What does a target audience see as attractive or unattractive features of a program?" Method: focus groups; unstructured or open-ended surveys and interviews	Question: "What is the effect of a program on specified attitudes or behavioral intentions?" Method: surveys; experiments; quasi-experiments	Question: "What is the effect of a proposed new program on enlistment?" Method: experiments; quasi-experiments
Existing programs	Same as above	Same as above	Question: "What is the effect of an existing program on enlistment?" Method: econometric modeling

mented without the use of systematic evaluation in their initial design and implementation. Chapter 5 offers further development and illustration of the use of such qualitative research methods in the context of developing advertising campaigns.

The framework identifies surveys as a useful method when the research question deals with the effect of existing or proposed programs on attitudinal or behavioral intent variables of interest. Survey questions can be administered in various media, from paper-and-pencil questionnaires to telephone interviews to administration via computer. The key distinction with the previous category is that the researcher has clearly specified in advance the attitudinal or behavioral intent variable of interest. Thus if a portfolio of possible new recruiting incentives is being considered, and the researcher is interested in the effects of each on an individual's stated likelihood of enlistment, the use of survey methods would be particularly appropriate. Note that experimental and quasi-experimental designs can be usefully employed in such settings, with different groups of participants asked to respond via survey to different proposed incentives. Survey methods are also useful in monitoring whether existing programs continue to generate the same attitudinal reaction over time.

It is often the case that the attitudinal or behavioral intent variables of interest are not of primary interest in and of themselves; they are of interest because of a documented link to actual behaviors of interest, such as enlistment. Because of this link to behavior, insight into the potential effects of programs can be obtained prior to actual program implementation. Thus, for example, a wide array of different bundles of recruiting incentives could be presented to research participants to obtain insight into which is seen as most attractive. Note, though, that while information on the relative attractiveness of different options can be quite useful in deciding which ones to implement or to research more extensively, this method does not result in a point estimate of the effects of the program on enlistment (without research linking survey responses to actual enlistment, as discussed in Chapter 7). Chapter 3 further develops and illustrates the use of survey methods.

The final two cells in the framework focus on behavior; in the military recruitment context, enlistment is the behavior of greatest interest. It is in research questions focusing on enlistment behavior that the distinction between the evaluation of new programs and the evaluation of existing programs becomes a critical issue. Econometric modeling methods are well suited to the evaluation of ongoing programs, as they are based on actual variations in a set of observable conditions and relating that variation to behavioral outcomes of interest. For example, the effects of advertising expenditures on enlistment can be examined by relating changes in expenditures over time to patterns of enlistment, statistically controlling for other relevant time-varying features. Chapters 5 and 6 further develop and illustrate econometric modeling.

In contrast, determining the effects on enlistment of a new program prior to full-scale implementation or the differential effects of alternate possible new programs are not amenable to examination via methods relying on observed variation in the features of interest, because the programs have not yet been implemented. Thus approaches involving manipulation of features of interest are needed, namely, experimentation (manipulation with random assignment) or quasi-experimentation (manipulation without random assignment). For example, different advertising content could be used in different geographic regions to examine effects on enlistment. Chapters 4 and 7 further develop and illustrate experimental and quasi-experimental approaches.

While the framework presented here identifies broad types of research questions for which particular research methods are well suited, it is not our intent to draw bright-line distinctions among the methods.

Another message of this volume is the relevance of multiple methods to many key issues in military recruiting. Chapters 6, 7, and 8 all illustrate areas in which multiple methods are useful. In addition, aspects of different methods can usefully be combined. For example, quasi-experimentation does not ensure the equivalence of conditions that are obtained via random assignment, and some of the quantitative tools of econometric modeling can be very useful in controlling for differences among conditions.

In the next chapter we elaborate more fully on similarities and differences between econometric and experimental approaches to research. Again, it is the nature of the research question that should drive the choice of research method or methods, rather than allegiance to any particular methodological orthodoxy.

2

Theoretical Approaches

As noted in Chapter 1, research is concept driven and shaped by the questions being asked and the variables being investigated. It is difficult to discuss research strategies for effectively designing and evaluating programs to increase enlistments without also considering theories about the core variables that impact enlistment decisions. Theories of enlistment behavior are diverse and have been influenced heavily by the disciplines of economics, sociology, and psychology. Not surprisingly, the kinds of variables emphasized and the research designs chosen to explore these variables differ somewhat from one discipline to another. For example, social-psychological theories of enlistment behavior tend to emphasize micro-level variables focused on characteristics of the individual and individual decision-making processes (relying on such constructs as beliefs, attitudes, perceived social pressures, and behavioral intentions), whereas economic theories tend to emphasize macro-level variables focused on such constructs as recruitment resources, the general state of the economy, wages, and work opportunities in military and civilian sectors. There is no single "correct" level of theorizing. Some theories are better poised to answer some questions than others and the level of theorizing is dictated, in part, by the particular question at hand.

When considering the impact of an incentive program or advertising campaign on enlistments, it is important to understand how that program or campaign affects key variables that govern the career choices of American youth. A major purpose of this chapter is to identify such variables. Our intent is not to present a comprehensive theory of enlistment behavior but rather to delimit the kinds of variables that program designers and advertisers should be thinking about as they develop campaigns to increase enlistments. More focused questions can be explored using the research strategies and techniques outlined in later chapters.

In its Phase I report, *Attitudes, Aptitudes, and Aspirations of American Youth*, the committee outlined a general theory of enlistment behavior (see National Research Council, 2003, Chapter 7). We begin by briefly reviewing this theory and then elaborate on it to incorporate additional perspectives from economics and research on adolescent development. We then describe how the identified variables can be incorporated into the design and evaluation of programs and advertising campaigns aimed at increasing enlistments, drawing on both the theoretical work reviewed as well as facets of communication theory.

THEORETICAL FRAMEWORK FOR ENLISTMENT BEHAVIOR

Perspectives from Behavioral Theory

The general theory offered by the committee in its previous report is reproduced in Figure 2-1. This integrative framework is based on several empirically supported theories of behavior and behavior change. These include the health belief model (Becker, 1974, 1988; Rosenstock, Strecher and Becker, 1994), the theory of reasoned action (Ajzen and Fishbein, 1980), social cognitive theory (Bandura, 1991, 1994), and the theory of planned behavior (Azjen, 1985, 1991). According to the framework, whether or not someone enlists in the military (the "behavior" variable in Figure 2-1) is a direct function of the person's intention to enlist: if a person does not intend to enlist in the military, he or she probably will not do so. If the person intends to enlist in the military, he or she probably will do so. The concept of intention maps roughly onto the construct of "propensity to enlist," a variable frequently encountered in the research literature on enlistments. The theory offered by the committee explicitly recognizes that intentions do not always translate into behavior. Sometimes people state a negative intention to enlist but, with time, end up enlisting in the military. Others fully intend to enlist in the military but fail to ever do so.

Two classes of variables affect whether a person's intention or propensity to enlist translates into enlistment behavior. One class of variables concerns environmental factors that either facilitate or prevent the person from carrying out his or her intention. Examples of environmental constraints include long distances from recruitment or enlistment centers, factors that preclude access to such centers, lack of recruiter activity, and lack of recruiter effort. Examples of facilitators include the presence of recruiters, high levels of recruiter effort, and different kinds of recruiter activities. The second class of variables that influences if intentions trans-

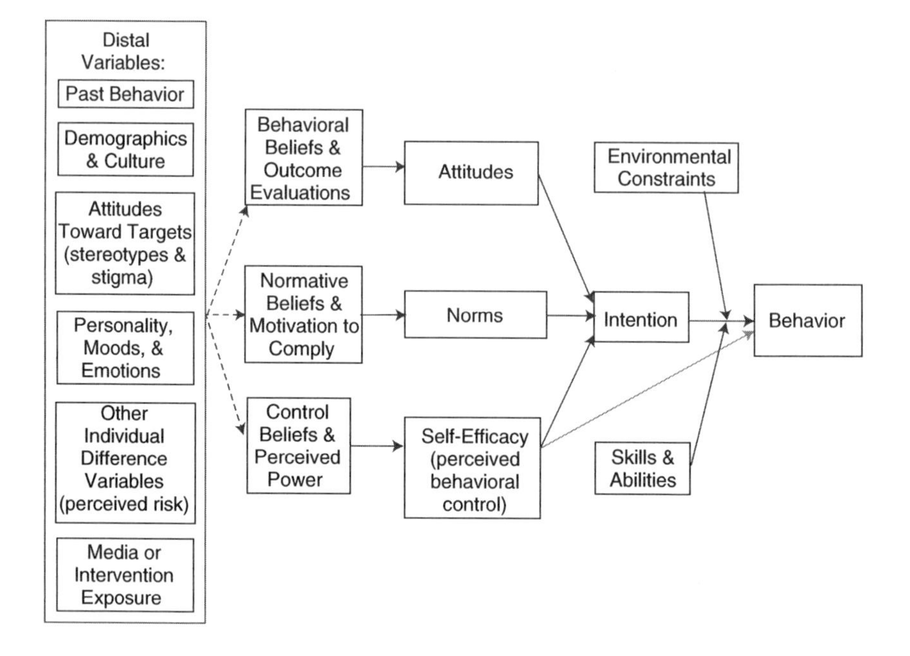

FIGURE 2-1 Determinants of behavior.
SOURCE: Reproduced from *Attitudes, Aptitudes, and Aspirations of American Youth.*

late into behavior is whether the individual has the requisite qualifications, skills, and abilities to perform the behavior. For example, a person may have a strong intention to enlist in the military, but if he or she cannot graduate from high school or cannot pass the requisite physical exam, then an enlistment will not result irrespective of the presence of a strong intention.

Despite these moderating influences, intentions to enlist tend to be good predictors of enlistment behavior (e.g., Bachman, Segal, Freedman-Doan, and O'Malley, 1998). Thus, it is of key interest to understand what factors cause some individuals to have positive intent with respect to enlisting in the military and others to have negative intent.

Determinants of Intentions to Enlist

According to Figure 2-1, there are three immediate determinants of a person's intention to enlist. The first is how favorable or unfavorable the individual feels about enlisting, that is, the individual's personal attitude toward enlisting. In general, the more favorable an individual feels about enlisting in the military, the more likely it is that he or she will intend to do so. Although many people base their decisions exclusively on such personal attitudes, others take into account what people important to them are doing and their perception of what those important others think they should do. For example, although a young man may have a highly favorable attitude toward enlisting in the military, he may not do so because of strong resistance and disapproval from his mother.

Thus, a second class of variables that can impact the intention to enlist is the normative pressures that are brought to bear on the individual. In general, the more a person sees important others as being supportive of a decision to enlist in the military, the more likely it is that the person will intend to enlist in the military. Such normative influences are the hallmark of many sociological theories of behavior (Bakken, 2002).

The third class of variables that influence intent concerns self-efficacy, that is, whether the individual believes he or she can, in fact, perform the behavior. Even though a person may have a positive personal attitude toward enlisting and even though important others may support that decision, if the person does not think he or she will be successful in attempts to join the military, then she or he will not even bother trying. These three variables—attitudes, norms, and self-efficacy—represent core influences on intentions to enlist (see Chapter 3 for additional discussion).

The model in Figure 2-1 also identifies the immediate determinants of each of these three variables. We now consider these determinants.

Determinants of Attitudes

Someone's personal attitude toward enlisting (i.e., how favorable or unfavorable the person feels about enlisting) is said to be a function of the individual's behavioral beliefs and outcome evaluations associated with those beliefs (Fishbein and Ajzen, 1975; Ajzen and Fishbein, 1980). To elaborate, individuals perceive certain advantages and disadvantages of enlisting in the military. A given perceived advantage or disadvantage has two components. First, there is an *expectancy,* which refers to how likely the individual thinks it is that enlisting in the military will, in fact, lead to the advantage or disadvantage in question. For example, a possible advantage of pursuing a career in the military might be that one will acquire a useful job skill for later in life. The expectancy in this case, also called a *behavioral belief,* is the subjective probability on the part of the individual that the person, in fact, will acquire a useful job skill for later in life if she or he enlists in the military.

The second component is an *outcome evaluation.* This refers to how positive or negative the advantage or disadvantage is perceived as being. Some advantages are thought to be more positive than others and some disadvantages are thought to be more negative than others. The individual's perception of the degree of positivity or negativity of a given consequence is also important to take into account. These perceptions derive from the individual's more fundamental value system, which ultimately determines the worth that he or she ascribes to the different outcomes and consequences.

Individuals perceive multiple advantages and disadvantages of enlisting. For each of *m* consequences potentially associated with enlisting, there will be a behavioral belief (i.e., subjective probability or expectancy) and an outcome evaluation. The overall attitude toward enlisting in the military is some function of these multiple expectancies and values:

$$\text{Attitude} = f\,(b_1, b_2, \ldots.\ b_m, e_1, e_2, \ldots\ e_m)$$

where Attitude is the overall favorability/unfavorability toward enlisting, *b* is a subjective probability that enlisting in the military will, in fact, lead to consequence *m*, and *e* is how positive or negative consequence *m* is perceived as being. Psychologists, sociologists, and decision theorists are in disagreement about the nature of the function relating behavioral beliefs and outcomes to attitudes (Anderson, 1996; Fishbein and Ajzen, 1975; Hastie and Dawes, 2001), but as a general rule, individuals will have more positive attitudes if they perceive enlisting in the military as definitely leading to highly positively consequences and definitely not leading to negative consequences. Individuals will have more negative attitudes if they perceive enlisting as definitely leading to highly negative consequences and definitely not leading to positive consequences.

Determinants of Normative Support

A second potentially relevant determinant of the intention to enlist is the social and normative pressure one feels to enlist or not to enlist. Two types of normative influence potentially contribute to this social pressure: injunctive norms and descriptive norms (Cialdini, 2003). Injunctive norms refer to perceptions of what important others think the individual should do with respect to enlisting. Descriptive norms refer to perceptions of base rates, or how many of one's peers are performing the behavior.

Injunctive norms reflect whether important others approve or disapprove of the individual's enlisting. According to Figure 2-1, perceptions of such approval or disapproval are reflective of or determined by the perceptions of the opinions of specific referents, such as one's mother, one's father, or one's boyfriend or girlfriend. There are multiple referents who may have an opinion about what the individual should do, and these referents may have conflicting opinions. The overall normative pressure to enlist or not enlist is some function of these differing opinions:

$$NP = f(NB_1, NB_2, \ldots. NB_k)$$

where NP is the overall normative pressure to enlist and NB_k is the strength of the opinion of referent k, as perceived by the individual, that the individual should or should not enlist in the military. The various NBs are called *normative beliefs* (see Figure 2-1).

Many theorists argue that it is important to take into account the individual's motivation to comply with a given referent (Fishbein and Ajzen, 1975). A relevant referent may have a strong opinion about what the individual should do, but if the individual has little motivation to comply with or please that referent, then the overall normative pressure felt by the individual will be lessened. Thus, we modify the above expression to reflect motivations to comply with referents, such that

$$NP = f(NB_1, NB_2, \ldots. NB_k, MC_1, MC_2, \ldots. MC_k),$$

where MC_k is the motivation to comply with referent k and all other terms are as previously defined. In general, people perceive a more supportive normative environment for enlisting if others who are important to them unanimously agree that they should pursue a career in the military and the individual is highly motivated to comply with those others.

The second type of normative influence, descriptive norms, has been identified as an important determinant of behavior in the literature on adolescent development (e.g., Borsari and Carey, 2003). Descriptive norms refer to perceptions of how many of one's peers are pursuing the choice option in question (e.g., most of my friends are enlisting, only a few of my friends are enlisting, or none of my friends is enlisting). There are different base rates for different referent groups. For example, the rate of enlistment might be different for one's close friends, one's general circle of friends, the people at one's high school, the people in one's community, and the nation as a whole. The overall base rate factor, BRF, is some function of these more specific group base rates:

$$BRF = f(BR_1, BR_2, \ldots. BR_k)$$

where BRF is the overall base rate factor for choosing a given career option, and BR_k is the perceived base rate of choosing the option for group k. The social psychological literature on base rates is complex: sometimes higher base rates lead to increases in behavioral intent, and sometimes lower base rates do. A useful theory for understanding the impact of base rates on behavior is deviance regulation theory (Blanton and Christie, 2003).

Determinants of Self-Efficacy

The third primary determinant of the intention to enlist is the extent to which a person feels he or she can be successful at enlisting given that the effort is made to do so (Ajzen, 1991). The primary determinants of global judgment of self-efficacy are perceptions of the obstacles that impede enlistment and one's judged ability (or perceived power) to overcome those obstacles. For example, an obstacle to enlisting might be that of obtaining a high school diploma, and an individual may be uncertain of his or ability to overcome this obstacle. Individuals may perceive multiple obstacles. Associated with each obstacle is a belief that the obstacle can be overcome. The overall judged self-efficacy is some function of these perceptions:

$$SE = f(O_1, O_2, \ldots O_n)$$

where SE is the overall judged efficacy for enlisting, and O is the judged likelihood of overcoming obstacle n.

In sum, each of the three core determinants of intentions to enlist have a set of immediate determinants themselves. The person's attitude toward enlisting is influenced by his or her behavioral beliefs and outcome evaluations, the person's overall subjective norm about enlisting is influenced by his or her perceptions of the opinions of specific referents and the motivation to please or comply with those referents, and the person's general feeling about self-efficacy is influenced by the perceived obstacles to behavioral performance and one's perceived ability (or power) to overcome those obstacles.

Distal Determinants of Enlistment Behavior

Although the above classes of variables are the most immediate determinants of enlistment intentions, many other variables influence enlistment decisions, as shown in Figure 2-1. The influence of these other variables is more "distal" in the sense that their effects on behavior are mediated by or reflected in the more immediate determinants. For example, feelings of patriotism may influence enlistment behavior, but any such influence should be traceable through the more immediate determinants outlined above. Perhaps those with higher levels of patriotism have more positive behavioral beliefs about enlisting than those with lower levels of patriotism. Or, perhaps those with higher levels of patriotism perceive a more supportive normative environment for enlisting than those with lower levels of patriotism. Or perhaps those with higher levels of patriotism are more confident in their abilities to enlist than those with lower levels of patriotism.

Figure 2-1 identifies five general classes of distal variables that may impact enlistment decisions and are important to take into account: (1) demographic and cultural variables (e.g., age, education, gender, socio-economic status, employment status, ethnicity), (2) attitudes toward targets (e.g., stereotypes associated with members of the military, social stigmas associated with being in the military), (3) personality, moods, and emotions (e.g., sensation-seeking propensities, positive mood states, gut-level emotional reactions to enlisting), (4) other individual difference variables (e.g., feelings of patriotism, risk-taking propensities), and (5) media or intervention exposure (e.g., exposure to positive or negative portrayals of the military in the media).

Perspectives from Economic Theories

The above theoretical framework is described using terms that derive from sociology and psychology. There are large literatures in economics at both the micro-economic (see for example, Asch and Warner, 2001; Dertouzos, 1985) and macro-economic (see for example, Daula and Smith, 1985; Warner, 1990; Berner and Daula, 1993; Murray and MacDonald, 1999) levels that provide additional perspectives on enlistment behavior over and above those elucidated in Figure 2-1. This section elaborates theoretical orientations from economics that complement and extend the above formulation.

Importance of Decision Options

Economic theories of enlistments often focus on individual decision making about whether to join the military or pursue a civilian alternative. Economic theory assumes that an individual who is considering military service makes the decision to join by comparing the perceived utility he or she expects to receive from military service (U_M) with the utility he or she will receive from remaining a civilian (U_C). The utility is the overall "worth" of a given career, taking into account all economic, social, and personal considerations.

In economic theories, it is not uncommon to elucidate the role of wages and compensation in influencing these utilities. For example, it might be asserted that the utility for a given career choice is based on the compensation associated with that choice (W) and other nonpecuniary aspects of that choice (γ). The utility associated with the military choice can be represented by the equation $U_M = W_M + \gamma_M$ and the utility associated with the civilian choice by $U_C = W_C + \gamma_C$. A youth prefers to join the military if the expected utility from military service exceeds the expected utility from civilian life, that is, if $U_M > U_C$ or, as simple algebra shows, if $W_M - W_C > \gamma_C - \gamma_M$.

A crucial feature of this simplified economic analysis is the idea that, to understand enlistment decisions, one must not only study how individuals construe the option of a military career but also how they construe competing career options in the civilian sector. It is only when a theorist considers both choice options that proper perspectives on enlistment behavior emerge.

The above economic theory conceptualized choice in terms of two decision options: (1) pursuing a military career or (2) remaining a civilian. However, classic decision theories in both economics and cognitive psychology do not preclude a more refined and detailed elaboration of choice alternatives. For example, when considering different military options, an individual might think about (1) enlisting in the Army, (2) enlisting in the Navy, (3) enlisting in the Air Force, (4) enlisting in the Marine Corps, (5) enlisting in the reserve forces, or (6) joining a college-based officer training program. Civilian alternatives might include (1) going to a two-year college, (2) going to a trade school, (3) going to a four-year college, (4) seeking a white-collar position in one's home town, and so on. An important task of decision analysis is determining what career options individuals perceive to be available to them and how they define and frame these options.

Optimizing and Satisficing

According to many economic choice theories, individuals associate an overall utility with each choice option and then choose to pursue the option that has the highest or most positive utility associated with it. This preference for the option with the most positive overall utility is called optimization. Consider four individuals, each of whom has the same four career options in mind, C_1, C_2, C_3, and C_4. Let C_1 represent the only enlistment option in the set of career options, and the remaining options represent civilian-sector options of one form or another. Suppose that the overall utility associated with an option is scaled from 0 to 1.00 with higher values indicating more positive overall utilities. The distribution of utilities for each individual might appear as follows:

Individual	C_1	C_2	C_3	C_4
1	0.90	0.03	0.01	0.01
2	0.48	0.17	0.12	0.14
3	0.48	0.48	0.30	0.30
4	0.17	0.17	0.33	0.48

The choice of each individual should be the alternative with the highest utility and the propensity to enlist or the strength of the decision to enlist is defined as the difference (or ratio, or some other function) between the utility associated with C_1 and the highest utility for the remaining options:

Individual	C_1	Highest Option	Difference
1	0.90	0.03	0.87
2	0.48	0.17	0.31
3	0.48	0.48	0.00
4	0.17	0.48	–0.31

The highest propensity to enlist is for individual 1 because she or he has a high overall utility associated with C_1 and a low utility associated with all remaining options. Note that even though individuals 2 and 3 have identical utilities associated with enlisting, they differ in their propensity to enlist because individual 2 does not have as viable alternatives to C_1 as individual 3. This example shows that it is the relative positioning of C_1 within the choice set that is crucial, so to understand behavior, one must study the entire choice set, not just one component of it.

As noted, economic theories of choice typically assume that individuals strive to optimize, that is, they choose options that have the highest overall utility associated with them. However, several theorists have argued that individuals sometimes use "satisficing" rather than optimizing strategies (Todd and Gigerenzer, 2003; Schwartz et al., 2002; Hastie and Dawes, 2001). In satisficing decision strategies, individuals set a minimum utility value that an option must surpass in order for that option to be deemed acceptable. If the overall utility of an option falls below this threshold, then it is rejected as a viable course of action. As opportunities for pursuing different career options occur over time, the individual chooses to pursue the first option encountered that meets or surpasses the minimum threshold. Such a satisficing rule can result in a career choice that is not optimal in the sense that the chosen option may not have the highest utility from those in the entire choice set. Rather, the career was the first "acceptable" option that came along.

Job opportunities often occur sequentially over time, and individuals may "satisfice" when an acceptable opportunity avails itself. This results in some people making choices for options that are actually inferior to later opportunities. An important contribution of economic theories of choice is the recognition of different functions relating option utilities to the choice of a given career.

Economic Conceptions of Utility

The concept of an overall utility for a choice option in economic theories roughly maps onto the concept of attitude in Figure 2-1. Economic theories also have parallels to behavioral beliefs and outcome evaluations. This is most apparent in multiattribute models of decision making, the classic example being that of subjective expected utility (SEU) theory (Von Neumann and Morgenstern, 1947; Hastie and Dawes, 2001). According to this theory, the different attributes associated with a career option (e.g., how much it pays, how much travel is involved with it, its social status) determines the overall option utility. More specifically, the overall utility is a function of the person's subjective probability that choosing the option will lead to each of the attributes or consequences in question (paralleling behavioral beliefs) and the utility or worth associated with each attribute or consequence (paralleling outcome evaluations).

Some economists argue that the other categories of influence in Figure 2-1 (norms and self-efficacy) are simply additional career attributes (not unlike wages, travel demands, and social status) that go into the calculus of determining the overall utility of a career option. As such, norms and self-efficacy do not deserve special status relative to other career attributes. By contrast, sociologists and psychologists argue that viewing the constructs of norms and self-efficacy in this way is too narrow, because these variables have special explanatory power and are of substantive interest in their own right. Just as economists want to build theories in which wages and compensation take on special roles, sociologists and psychologists want to build theories in which constructs like norms and self-efficacy take on special roles.

Economic models provide conceptual strategies for thinking about how specific attributes associated with a career option impact choice. This can be illustrated using the analysis of wage differentials described earlier. Recall that the overall utility associated with the military choice was defined as $U_M = W_M + \gamma_M$ and the utility associated with the civilian choice is $U_C = W_C + \gamma_C$ (where W is the compensation associated with a given option and γ are nonpecuniary aspects of the option). Based on the optimizing principle, an individual is said to prefer to join the military if $U_M > U_C$. Simple algebra shows that an individual prefers military service if $W_M - W_C > \gamma_C - \gamma_M$. Thus, in order for individuals to choose to enlist, the military-civilian wage differential (W_M – W_C) must exceed the *net* value the individual places on the nonpecuniary aspects of civilian life ($\Delta = \gamma_C - \gamma_M$). An increase in the wage differential $W_M - W_C$ increases the likelihood that a youth will prefer military service, but an increase in the value the youth places on the net value of civilian life reduces it.

The enlistment rate in the population is the fraction of youth for whom $W_M - W_C > \Delta$. If the net preference factor Δ is completely random, then the enlistment rate depends only on military and civilian compensation. It is more likely, however, that the net value that a youth places on the nonpecuniary aspects of civilian life, Δ, depends on social and environmental background factors. Youth no doubt acquire attitudes about military service from parents, relatives, and friends during their formative years. To the extent that they form better impressions about the military and the importance of service during this formative period, the higher γ_M is and the more likely it is that $W_M - W_C > \Delta$.

The above example illustrates a classic economic strategy for thinking about and conceptualizing the impact of a given attribute in a task involving choice. Economic models thus offer tools and strategies for conceptualizing the impact of career attributes on career choice that are somewhat different, yet complementary, to the approach based on Figure 2-1.

Recruiter Activity and Recruiter Effort

Economic models of enlistment behavior often emphasize the central role of recruiter activity and recruiter effort in influencing enlistments (see Chapter 5 for a review of this research). Such variables are either distal variables in the model in Figure 2-1 or environmental facilitators that help individuals translate positive intentions into behavior. Economic models capture recruiter dynamics in ways that are not elaborated in traditional theories using the constructs of Figure 2-1 and hence complement the framework of the figure in important ways.

Economic models recognize that recruiters are not randomly distributed across recruiting areas, but rather are concentrated in areas in which there is naturally more fertile ground for prospects. Because of this, simple cross-sectional estimates can provide biased effects of recruiter productivity. In addition, economic models recognize that recruiters are provided with quotas or goals. Analysis of recruiting behavior suggests that failure to account for goals in model estimation may result in biased estimates of recruiter productivity and biased estimates of other factors affecting recruiting, such as enlistment bonuses. Recruiters' preferences, recruiting technology, and recruiter incentives also are of central concern in economic models.

A review of relevant economic research on recruiter effects is presented in Chapter 5. These economic modeling efforts are important because they elaborate the role of the recruiter in enlistment decisions and enlistment behavior.

Other Distal Variables

Economic models of enlistment behavior focus on many distal variables that are represented only abstractly on the left-hand side of Figure 2-1. The particular variables studied in a given economic model depend on the specific questions being addressed. Although it is not common, it is possible to integrate the analysis of such economic variables with the more immediate, social-psychological behavioral determinants in Figure 2-1.

Consider, for example, a variable such as the state of the economy. A poor economy influences the range of job opportunities an individual has available, that is, it influences the set of career options that the individual chooses among. In a poor economy, a military career option competes with fewer civilian alternatives, and the intention to enlist is framed in terms of this more restricted set of options. In a poor economy, there may be reductions in the level of funding available for recruiter activity and outreach. This can impact whether a person's intention to enlist is translated into behavior by removing "environmental facilitators" that reduce the hassles and obstacles to actually enlisting (see the category in Figure 2-1 "Environmental Constraints"). A poor economy also may influence the kinds of behavioral beliefs that an individual takes into account when evaluating different options in the choice set. Issues of salary and benefit packages may take on increased importance in the cognitive calculus of the individual as he or she weighs the advantages and disadvantages of different career options (see the category "Behavioral Beliefs and Outcome Evaluation" in Figure 2-1). In addition, a poor economy also may impact the kind of normative pressures that are brought to bear on the individual. For example, a parent or spouse who may have been more tolerant of the individual's pursuing a military career may be less so if that parent or spouse sees a need to generate greater and more immediate income.

Although there are other ways in which the effects of a poor economy can manifest itself in the theoretical system, our general point is that an integrated analysis of economic, sociological, and psychological variables will have substantial payoffs in providing insights into the mechanisms by which the different variables impact enlistment behavior. The framework of Figure 2-1, coupled with its elaboration based on economic models of enlistment behavior, provides a blueprint for researchers and program planners when pursuing such integrated efforts. This blueprint can be augmented by additional variables that are specific to the more immediate, practical questions being addressed.

Summary

Economic models offer additional conceptual tools and foci that complement the framework in Figure 2-1 and that facilitate development and evaluation of interventions, advertising campaigns, and general policy setting for increasing enlistments. These foci include (1) thinking about how an intervention or campaign affects the placement or positioning of a military career to other career options that an individual might be contemplating, (2) recognizing that individuals do not always optimize their choices but instead may invoke simplifying heuristics that result in some form of satisficing, (3) thinking about the concept of an overall utility associated with a choice option and how individual attributes associated with a career impact that utility, (4) recognizing the important role of recruiters and how they impact the various components outlined in Figure 2-1, and (5) calling attention to important economic variables that are distal determinants of enlistment behavior and whose effects can be traced through the more immediate determinants outlined in Figure 2-1.

Perspectives from Adolescent Development

For large segments of the population, enlistment decisions are made during late adolescence. There are research literatures in adolescent development suggesting that program designers and advertisers should consider at least one construct not elaborated in Figure 2-1, namely self-concept. Adolescents tend to be concerned about the images they project to others. Adolescence also is a time when youth are actively involved in identity formation. Adolescents want to carve out and transition to an adult identity that they can embrace and that is positively viewed by others. Such self-images represent distal variables in Figure 2-1, but they are so central to late adolescence that they should at least be considered (even if dismissed) when thinking about career choices that adolescents make. The choice of a career has major implications for one's self-image and the image that one projects. It follows that such variables will be of potential relevance for enlistment behaviors.

Self-concept can be conceptualized usefully based on the framework of social prototypes (Gerrard et al., 2002; Thornton, Gibbons and Gerrard, 2002). Social prototypes refer to images that individuals have of the kind of person who pursues a given choice option (e.g., the image of the kind of person who enlists in the military). Of interest is how positively this image is perceived as being and the extent to which a person's own self-concept maps onto the prototype of the kind of person who enlists in the military. If there is a close match between the person's self-concept and the prototype of the person who chooses the career option, and if the prototype is positive in character, then the individual will be more likely to pursue that option, everything else being equal.

The similarity between an individual's self-conception and the conception of a social prototype on a given dimension can be defined as $S_k - P_k$, where S_k is the extent to which the individual believes he or she is characterized by attribute k and P_k is the extent to which the social prototype is characterized by attribute k. (Note: other functions than a difference function could be used to represent this discrepancy.) U_k refers to the utility of attribute k, or the extent to which it is perceived as being positive or negative in character. The overall social prototype factor is some function of these discrepancies and utilities:

$$SP_j = f\,(S_1 - P_1, S_2 - P_2, \ldots, S_k - P_k, U_1, U_2, \ldots, U_k)$$

where SP_j is the overall social prototype factor for choosing career option j, and all other terms are as previously described. In general, individuals will be more likely to say they will pursue an option if their image of themselves maps onto the social prototype of the kind of individual who pursues the career option on dimensions that are positive in character.

In sum, research on adolescent development suggests that self-image (and the notion of prototype) should be considered when thinking about intervention strategies to increase enlistments.

Rational, Irrational, and Emotion-Driven Decisions

A common complaint about economic and social-psychological models of decision making is that they assume individuals are "rational" decision makers who carefully weigh the "costs" and "benefits" of their actions when making choices. The argument is that decisions do not always reflect rational information-processing strategies but instead are governed by irrationality, emotions, and impulsive tendencies (Steinberg, 2003). A strength of the framework in Figure 2-1 is that it can accommodate both perspectives.

From an information-processing perspective, the framework assumes that individuals take into account the *perceived* costs and benefits of the different courses of action available to them. These perceptions may be true or false in reality. For example, a person may think a certain civilian career will pay more than a military career when this is not, in fact, the case. Individuals act on their perceptions independent of their actual truth value. Thus, what may appear as irrational behavior to an observer who knows the true facts may be quite rational to the individual decision maker, who is maximizing utilities *as he or she perceives them.* By applying the framework, insights can be gained into misperceptions that individual decision makers have as they think about military career options.

The model in Figure 2-1 also recognizes that some individuals do not even consider the personal costs and benefits that will accrue to them if they make certain choices. Such behavioral beliefs are only one of three major variable categories that can impact enlistment decisions, the other two being normative pressures and self-efficacy considerations. As noted earlier, the relative impact of each of these variable categories on enlistment decisions can vary as a function of individual differences, with some individuals completely ignoring one class of variables in favor of another and other individuals taking into account different combinations of variable categories. In this sense, the framework goes well beyond simple rational, cost-benefit models of choice, allowing for the possibility that costs and benefits are not considered at all.

Finally, the framework recognizes that emotion and impulse-based variables can influence enlistment behavior. Emotions are distal variables that can shape what behavioral beliefs, norms, and control beliefs are taken into account when making decisions (see the box in Figure 2-1 labeled "Personality, Moods, and Emotions"). In addition, impulse and emotional control represent a "skill and ability" that one must have in order to ensure that a chosen course of action is translated into behavior.

In sum, the framework in Figure 2-1 allows for formal information-processing determinants of enlistment behavior as well as irrational, emotional, and impulse-based determinants.

IMPLICATIONS FOR INTERVENTION DESIGN

The design of programs or advertising strategies to increase enlistments will be more successful if they are informed by research that identifies the core factors that are driving the career choices of a given target population. Once these factors are known, the campaign can attempt to directly change those factors or mitigate or enhance their influence. In general, interventions will tend to be more effective if they target the more immediate determinants of behavior as opposed to more distal determinants.

Figure 2-1 and the additional perspectives derived from economics and adolescent development suggest the following strategies for increasing enlistments:

1. Help individuals translate positive intentions into behavior by removing environmental constraints to enlisting and putting into place environmental facilitators.

2. Help individuals achieve the qualifications and skills necessary to translate positive intent into behavior (e.g., through educational assistance programs).

3. Change personal attitudes toward enlisting by (a) changing the subjective probability associated with a given consequence of enlisting (e.g., "you think you might learn an important job skill but I can assure you that you definitely will"), (b) changing the outcome evaluation or utility associated with a given consequence (e.g. "having this job skill is a really desirable thing for you to have and you don't realize just how valuable it will be"), or (c) adding a consequence to a person's belief system that they had not thought about before.

4. Change the overall normative pressure to enlist by changing either the injunctive or descriptive norms. With respect to the former, change can be brought about by (a) changing the referents' opinions about what the individual should do (e.g., by targeting parents of potential recruits as part of the campaign), (b) changing the individual's *perception* of a referent's opinion, (c) rendering the referent irrelevant to the decision in the eyes of the potential recruit, or (d) making the opinion of a supportive referent more important and salient to the decision maker. With respect to the latter, change can be brought about by (a) educating individuals about the true base rates of enlisting for different referent groups or (b) rendering a referent group to be more or less important to the individual.

5. Change the overall perceived self-efficacy associated with enlisting by (a) convincing the individual that a perceived obstacle is not an obstacle after all or (b) convincing the individual that he or she has the skills and wherewithal to overcome the obstacle.

6. Change the relevant social prototype and self-image by (a) changing how one perceives the kind of person who enlists in the military on a given attribute dimension (e.g., "people who enlist are patriotic"), (b) changing the utility or evaluation of an attribute dimension associated with the prototype (e.g., "being patriotic is a noble and highly desirable character quality"), (c) making new attribute dimensions associated with the prototype salient to the individual, or (d) changing how one perceives oneself on one or more of the attribute dimensions associated with the prototype.

7. Alter any of the above for competing options in the choice set so that the options associated with a military career rise above those of their civilian competitors in terms of overall utility.

8. Introduce new military-based career options to the choice set that the individual may not have thought about and that will be relatively attractive to the individual.

9. Alter any of the above through recruiter activities and programs aimed at increasing the effectiveness of recruiters.

Some of these strategies will be more effective in some target populations than others. The committee in its previous report stressed the idea that the impact of any single variable in Figure 2-1 can vary from population to population. A behavioral belief that might be highly relevant to male recruits may be irrelevant and inconsequential to female recruits. The impact of injunctive norms for one's parents might be stronger for some ethnic groups than others. An important step in program design is conducting the requisite research to isolate which of the many possible determinants of enlistment behavior are likely to have the biggest effect if changed. Based on this research, effective programs can be structured.

IMPLICATIONS FOR INTERVENTION EVALUATION, LONGITUDINAL ANALYSIS, AND DEMOGRAPHIC ANALYSIS

When a campaign or program to increase enlistments has been put in place, the above theoretical framework can be used to help evaluate why the program works, does not work, or ways that the program might be improved to enhance its effectiveness. As an example, consider a program designed to make a military career more attractive by offering a free laptop computer to those who enlist. The idea of the program is to add a new behavioral belief about a positive consequence that will occur if the person enlists and this, in turn, should impact enlistment behavior. The above theoretical framework suggests a number of strong assumptions that such a program hinges on. First, one must assume that having a free laptop computer is evaluated sufficiently positively by individuals to result in a nontrivial change in personal attitudes toward enlisting. Second, one must assume that personal attitudes toward enlisting are a significant determinant of the intention to enlist. If normative or self-efficacy factors are the primary determinants of intention for the target population rather than attitudes, then the program may not be successful even if it affects personal attitudes. Third, one must assume that the changes in personal attitudes that result are sufficient to make those attitudes more positive than the attitudes toward the other civilian options in the choice set. Even if attitudes toward enlisting become more positive, if individuals still feel more positive about a civilian alternative, the program will be ineffective in increasing enlistments. Fourth, even if the program produces a change in the intention to enlist, this still may not result in increased enlistments unless those intentions translate themselves into behavior. To the extent that environmental constraints mitigate against doing so, the program will be ineffective in increasing enlistments.

All of the above mechanisms can be measured directly in evaluation research, permitting analysts to pinpoint if the program has achieved its desired effects and, if not, why not. Alterations to the program can then be enacted to increase its effectiveness. For example, if the problem is that the laptop being offered does not produce a sufficient change in personal attitude, then perhaps offering a higher quality laptop will increase its attractiveness enough to change the attitude. If the problem is that the change in attitude is not sufficient to produce a change in intent, then one might add additional incentives to complement those of the free laptop, making the attitude toward enlisting that much more positive. If the change in attitude fails to lead to a change in intent because the intention is primarily influenced by injunctive norms, then the program might be augmented to ensure that parents of the recruit learn about the awarding of a laptop, thereby reducing their resistance to their son's or daughter's enlisting. Or if the program results in a change in intention but not behavior, then the program might be enhanced to address the environmental constraints that are preventing the recruit from translating his or her positive intent into behavior.

Evaluation efforts are more informative not only if they address the effectiveness of a program in changing the behavior of interest, but also if they provide perspectives on why a program fails to work or how that program might be improved. Applying the theoretical perspectives described in this chapter to evaluation research can help achieve these goals.

The above logic also applies to understanding the longitudinal dynamics of enlistment behavior and how and why enlistment rates change over time. In the committee's earlier report, it was noted that there was a decline in the propensity to enlist during the early 1990s and that this decline predicted the enlistment shortfalls in the late 1990s. During this time period, many distal variables were changing as a result of the economic dynamics at work in the United States: military pay relative to civilian pay declined, the economy boomed, and the investment in recruiting efforts shrank. As these distal economic variables changed, their impact on enlistment behavior should have been reflected in concomitant (but perhaps lagged) changes in the mediating variables of Figure 2-1.

For example, the decreased pay for military careers might have changed behavioral beliefs about the salary associated with the military, which in turn would negatively impact the attitude toward enlisting. Lower pay structures also might have impacted the kinds of normative pressures that were brought to bear, as important referents (e.g., parents) discouraged the individual from pursuing a career that they thought would result in financial strains in the long run. Lower investments in recruiting activities might have lessened the facilitating influences that recruiters had in helping recruits translate positive intentions into behavior. By carefully analyzing how the mediators of Figure 2-1 change in conjunction with changes in broad economic and societal variables, one can gain insights into the mechanisms by which those changes are impacting enlistment behavior.

Finally, the same logic can be applied when analyzing the effects of any distal variable on enlistment behavior. For example, suppose that there are differences in enlistment rates for two different ethnic groups. To gain insights into the bases for these ethnic differences, one can compare groups on any of the core mediators in Figure 2-1. Group differences in enlistment behavior should be reflected in group differences in one or more of these variables. For example, it might be found that members of one group tend to perceive military careers as offering more opportunities for advancement than members of another group (a behavioral belief). Or it might be found that the parents of members of one group tend to be more approving of a military career than the parents of members of the other group (a normative belief). Integrative analyses that include both social-psychological variables and a theoretically driven set of distal variables can be more informative than studying either group of variables alone.

IMPORTANCE OF COMMUNICATION THEORY

The first step in designing an effective campaign to increase enlistments is to identify the factors that are most influential in determining enlistment behavior. Once identified, the program designer must then develop a strategy for changing, mitigating the impact of, or enhancing the impact of those determinants. The theoretical framework presented above can help identify relevant behavioral determinants, but it says little about how to change the core perceptions that underlie enlistment behavior. For example, how does one go about changing a specific behavioral belief? How does one change the outcome evaluation or utility of an attribute? How does one change the perception of an injunctive norm? How does one increase the impact of an attitude?

There are a wide range of strategies that might be used in such efforts, but one of the most common strategies is to provide individuals with new information that they have not previously considered. Providing information requires effectively communicating that information to recruits, thus bringing to bear the importance of communication mechanisms, a topic we now consider.

Classic conceptualizations of communication distinguish among five components in the communication process: (1) the source of a communication, (2) the communication itself (often referred to as the message), (3) the medium or channel through which the message is transmitted (e.g., face-to-face, written materials, visual recordings), (4) the recipient or audience of the communication, and (5) the context in which the communication occurs (McGuire, 1985). Each of these components of communication has subcomponents. For example, sources of a message may differ in their age, gender, expertise, and trustworthiness. Recipients of communications differ in terms of their motivational states, their emotional states, their past experiences, and their expectations. The surrounding environment varies in terms of its temporal, physical, social, and cultural features. Variations in each of these five factors represent independent variables that ultimately may affect the behavior of the potential recruit in response to the communication. The impact of a message about enlisting may vary as a function of the characteristics of the person delivering the message, characteristics of the message itself, characteristics of the channel through which the message is delivered, characteristics of the potential recruit, and characteristics of the context in which the communication occurs. It is likely that these variables interact in complex ways to affect responses to communications.

Communication also involves cognitive processes that can be affected by the above independent variables. For a communication to have meaningful impact on potential recruits, the recruits must first be exposed to the communication, they must attend to the communication, they must comprehend the communication (i.e., the meanings intended by the source must map onto the extracted meanings of the recruits), they must accept the extracted meanings as being valid, and they must retain these meanings in memory. At later points in time, these meanings may need to be accessed from memory, thereby invoking fundamental processes of retrieval.

The processes of exposure, attention, comprehension, acceptance, retention, and accurate retrieval form the foundation for meaningful communication. The processes are intertwined in the sense that each usually must occur for nontrivial communication to result. If the probability of exposure is 0, then meaningful communication will not result. If the probability of message comprehension is 0, then meaningful communication will not result. Even when the probabilities associated with each process are large, the probability of overall meaningful communication is attenuated. For example, if the probability of exposure, *p(Ex)*, is .90, the probability of attention, *p(At)*, is .90, the probability of comprehension, *p(Co)* is .90, the probability of acceptance, *p(Ac)* is .90, the probability of retention, *p(Re)*, is .90, and the probability of accurate retrieval, *p(Ret)*, is .90, and if these probabilities follow a simple multiplicative combinatorial rule, then the probability of meaningful communication is *p(Ex) p(At) p(Co) p(Ac) p(Re) p(Ret)*, which is $.90^6 = .53$, or about 50-50.

This simplified cognitive analysis underscores the complexity and challenges for fostering effective communication between change agents and potential recruits. The complexity is magnified when one realizes that the five facets of communication identified earlier (source, message, channel, recipient, and context) can affect each of the six cognitive processes differently (as main effects or in complex interaction with one another), that recruits are often exposed to multiple and sometimes conflicting communications, and that change agents also must accurately interpret the communications of recruits in the context of reciprocal communication. Note also that none of this bears on the effect of the communication on enlistment behavior, which is the ultimate criterion for the intervention effort. Although a change agent may have a well-thought-out meaning structure to communicate to a recruit and although the change agent may believe that effectively communicating this meaning structure will increase the likelihood of enlistment, it is possible that the change agent may be wrong in that the ultimate acceptance, retention, and retrieval of the meaning structure is irrelevant to enlistment behavior.

This brief review of communication theory underscores an additional set of variables that program designers must take into account. To the extent that a program involves the communication of information to the recruit, program designers need to think about who the source of the communication will be, what the message will look like, the nature of the audience who is likely to be receiving and processing the message, the context in which the message will be delivered and subsequently processed, and the channel or medium over which the communication will be conveyed. Choices concerning the above should be enacted so as to maximize exposure to the message, attention to the message, comprehension of the message, acceptance of the message, retention of the message, and accurate retrieval of the message from memory. There exist large bodies of research in communications, psychology, and sociology to help address such issues.

CONCLUSION

The role of theory is crucial to the design of interventions to increase enlistment behavior. When such programs are developed atheoretically, they run a great risk of being ineffective. In this chapter, we have outlined a general framework for thinking about effective recruiting program design. The first step is to identify the fundamental factors that impact a target population's enlistment behavior. The second step is to derive strategies (often informationally based) to change, enhance the effect of, or mitigate the effect of those determinants. We outlined a wide range of variables and processes that program designers must potentially take into account, drawing heavily on research from adolescent development, communications, economics, psychology, and sociology. These perspectives set the stage for conducting the necessary research to inform program design and program evaluation.

3

Monitoring Trends in Youth Attitudes, Values, and Propensity

In our earlier report (National Research Council, 2003), we argued that "military readiness may best be served when the first role of military advertising is to support the overall propensity to enlist in the youth population and to maintain a propensity level that will enable productivity in military recruiting" (p. 6). On this basis, we recommended that "a key objective of the Office of the Secretary of Defense should be to increase the propensity to enlist in the youth population" (p. 8).

Clearly the more one knows about the determinants of propensity, the more it becomes possible to develop effective communications or other types of interventions to increase propensity. In this chapter we propose a cohort-based sequential sample survey with a longitudinal component that we think will provide the information necessary for both tracking the determinants of propensity and for developing more effective communications to increase and maintain the pool of youth with the propensity to join the military. This survey design has several advantages over current and recent survey designs that have been used by the Department of Defense (DoD) to track youth attitudes toward military service:

1. By assessing an unchanging set of core questions annually, this survey design provides a consistent set of data that can be used to monitor and track changes in the determinants of propensity, as well as in propensity itself, over time. Consistent measures from a constant set of questions are obviously necessary to discern changes in propensity over time across different cohorts. This perhaps can provide an early warning to changes in the recruiting market and provide some indication regard-

ing how advertising and other resources and incentives should be adjusted to reflect market changes.

2. One can assess the reasons for differences (cross-section) and changes (longitudinal) in key outcome variables over time. With structured modeling, the Services can determine which of the underlying determinants of propensity are driving current levels of propensity, and how they may have differed in the past. This will affect the development of advertising messages and the levels and types of incentives used to attract youth into military service.

3. The longitudinal nature of the data—in particular, the follow-up of individuals—provides the basis for assessing how critical determinants of propensity evolve over time. Not only will this allow the assessment of changes in the relative importance of factors influencing propensity to enlist over time, but also the follow-up sample will permit analyses of the relationships among initial intention, subsequent intention, and actual status (enlistment, college attendance, or entering the civilian workforce). These latter analyses may suggest ways to help young adults act on their initial intentions to enlist.

In the previous chapter, we proposed a model that suggested that propensity to join the military (or a particular Service such as the Army or Navy) was primarily determined by attitudes, norms, and self-efficacy (see also National Research Council, 2003). According to the model, the relative importance of these three variables as determinants of intention can vary as a function of both the behavior and the population being considered. That is, some behaviors may be primarily influenced by attitudes while others may be primarily influenced by norms or self-efficacy. Similarly, a behavior that is primarily under attitudinal influence in one population (or population segment) may be primarily under the influence of norms or self-efficacy in another population.

In order to develop interventions to increase the proportion of the population with a propensity to enlist at any given point in time, one needs to assess the underlying determinants of propensity. That is, the more one knows the values and relative importance of attitudes, norms, and self-efficacy in a given population, and the more one knows about the beliefs underlying these attitudes, norms, and perceptions of self-efficacy, the more likely it is that one can design an effective message (or other type of intervention) to increase propensity (Fishbein et al., 2001). However, this type of information (i.e., the most relevant data for guiding the development of effective messages or other interventions to increase propensity) is currently not available. Although some of this information was assessed by the Youth Attitude Tracking Study (YATS), a major problem with YATS is that, rather than getting complete data from each respondent, respondents answered only small, randomly selected subsets of the questions, making complete analyses at the individual level impossible.

It was for this reason that the committee prepared a letter report evaluating YATS, which was included as Appendix A in the committee's Phase I report (National Research Council, 2003), and presenting a number of findings and recommendations. Among these was the recommendation that "ongoing surveys to assess the critical determinants of propensity should be conducted on a regular basis." These surveys would allow for individual-level analyses to identify the psychosocial determinants of propensity. Thus, we recommended that, whenever a survey is designed, "consideration should be given to randomly assigning interrelated blocks of information to the same subgroups. Consideration should be given to maintaining sufficient sample size and content within a block of relevant questions so that multivariate analysis can be conducted without serious missing data problems" (p. 294).

In addition to conducting a survey to identify the determinants of propensity, it is important to assess changes in these underlying beliefs over time. Thus we also recommended that "DoD consider using a continuous tracking survey methodology for such issues as propensity to enlist, advertising awareness, awareness of direct response campaigns, involvement in high school activities, and perceptions of the military" (p. 295). More specifically, we recommended that "a portfolio of surveys at different time intervals replace the current annual YATS administration" (p. 294). Moreover, we suggested that, to be maximally useful, the formatting of the questionnaire must permit individual-level multivariate analyses. For example, it should permit assessment of an individual's complete set of salient beliefs about the consequences (or costs and benefits) of joining the military.

It is important to note that in response to these suggestions, DoD funded Wirthlin Worldwide to conduct a number of youth and influencer surveys (Bailey et al., 2002; Sattar et al., 2002). More specifically, Wirthlin conducted four surveys with respondents between ages 15 and 21 (March and April 2001; July and August 2001; October and November 2001; and October and November 2002) and four surveys with adults ages 22 to 85 (May 2001; September and October 2001; January 2002; and September 2002). While the youth surveys provided information necessary for testing some of the relations in the theoretical model, no single survey obtained information on all of the theoretical determinants of propensity. Thus while it was possible to examine some relationships that could not be examined by the YATS data, a full test of the model was not possible. Moreover, a large number of items focused on "generational" questions and in particular on variables assumed to be related to "millennials" (e.g., on team orientation, decision making, life satisfaction). There is little evidence that variables of this type are in any way related to the propensity to enlist or to actual enlistment behavior (National Research Council, 2002).

Turning to the adult surveys, Wirthlin does provide some interesting and important data concerning adult perceptions of the military and adults' propensity (or intentions) to "encourage [young people] to join a military service" or to "get a job" or to "attend a four-year college." It is interesting to note, however, that there are no questions assessing the determinants of these intentions to encourage youth to choose any of these career paths.

In sum, the current set of surveys does not yet provide the data necessary for identifying the critical determinants of propensity, and thus they do not provide the data necessary for developing effective interventions (including mass media advertising) to increase the pool of individuals with a propensity to enlist. In order to assist and provide guidance for mass media and other interventions to increase this pool of people, complete data concerning the determinants of propensity at the individual level are needed. Data are also needed to evaluate the effectiveness of advertising and other interventions. Thus, in addition to assessing whether a communication campaign has produced changes in propensity (and its underlying determinants), it is also necessary to track the extent to which the message is reaching its intended audiences (i.e., it is important to assess exposure to message content). In addition, in order to evaluate message (or advertising) effectiveness, it is necessary not only to determine whether a message is having a direct effect on propensity and its determinants, but also to look for other paths of effect—for example, messages can influence relevant referent (such as parents) who, in turn, influence youth; messages may increase conversations among youth vis-à-vis joining the service; messages may increase the likelihood that youth will talk to others (people in the military, school counselors, parents) about joining the military. One implication of these other paths is that, although a campaign may be effective, one may not always find differences between those who are and those who are not exposed to the advertising campaign. Thus, surveys should be designed to take these other paths of effect into account.

USE OF SURVEY RESEARCH

Survey research can be used for many purposes. A survey conducted at a given point in time can provide important prevalence data about propensity and its psychosocial determinants. These data also permit the investigation of relationships among beliefs, attitudes, perceived norms, self-efficacy, and propensity. Equally important, experimental or quasi-experimental designs or both can be built into a given survey. For example, in order to investigate how changes in one or more variables (or policies) could affect propensity, respondents taking the survey could be randomly assigned to different forms of the survey instrument that presented different recruitment or enlistment scenarios. Given this manipulation, one could then assess, for example, how different increases in pay or different education policies influenced the likelihood that one would join the military. Furthermore, by repeating the survey (or parts of it) over time, either cross-sectionally or longitudinally, one can observe changes in any of these variables or in the relationships among them. Moreover, by repeating surveys one can develop quasi-experimental (time series) designs to assess the efficacy of a given advertising campaign or recruitment strategy as a means of producing changes in propensity and its underlying determinants (Campbell and Stanley, 1966; Shadish, Cook, and Campbell, 2002).

Even when the major goal of conducting a survey is to provide information to increase propensity to join the military, there are a number of issues that must be addressed prior to developing one or more survey instruments.

1. What is the target population? What age group (or groups) should be targeted? That is, should one attempt to increase propensity among youth, adolescents, or young adults? Alternatively, knowing that there are important others who may strongly influence a person's propensity to join the military, one must decide whether to view these influencers as the target audience. It is therefore important to consider whether different surveys are needed for different populations and, if so, how often should each be conducted.

2. With one or more target populations identified, one needs to consider a number of sampling issues: size of sample(s), number of (distinguishable) samples, response rate, and other biases. In addition, one should consider questions of timing (annual, quarterly, continuous).

3. What survey methodology should be used and where and how can one maximize access to the population of interest? More specifically, should the mode of administration be face-to-face, self-administered, conducted by telephone (each of which can also be computer assisted), conducted via the Internet, or should it be a mail questionnaire? Similarly, should it be individually or group administered (e.g., at school)?

4. Turning to the content of the survey, one must consider such things as question ordering, question coverage, and question wording. Other issues relate to question types; for example, should the questions be open-ended (with and without prompts) or closed-ended (multiple choice)? What should be the order of presentation of the questions?

The remainder of this chapter will discuss issues related to survey structure, implementation, and content.

SURVEY STRUCTURE AND IMPLEMENTATION DECISIONS

The issues and recommendations presented here are broadly compatible with those in our 2000 letter report concerning YATS, but here we provide a greater degree of specificity along some dimensions. What is recommended here need not be viewed as a proposed replacement for the current pattern of telephone surveys, except insofar as cost constraints dictate some degree of trade-off. That said, it should be acknowledged that if the series of monitoring surveys recommended here were to be carried out, the overlap in coverage is such that it would eventually be possible to phase out much of the telephone survey work now being done.

The focus of this section is on monitoring trends in youth attitudes, values, and propensity. Nevertheless, the survey strategies outlined here could readily be adapted to incorporate other material. One example would be to measure, and perhaps monitor, reactions to advertising (see Chapter 5), although we would not propose this to the exclusion of other survey and nonsurvey methods focusing on advertising. Another example, as suggested above, would be to explore reactions to possible new or modified incentives for military service (see also Chapter 7).

A number of matters will have to be resolved before undertaking a set of surveys such as those proposed here. We outline them briefly, and in each case state the committee's recommendations.

Need for Long-Term Funding

Monitoring survey designs are built with the expectation that funding will be available to carry out the research over a considerable period of time. It is not a good investment of federal funds—or of investigator time—to initiate monitoring survey efforts only to drop funding after a short period of time. Careful consideration needs to be given to the sponsor's ability and willingness to make funding commitments for multiyear (e.g., five-year) periods. The survey approach outlined here supposes such funding commitments, and the committee strongly urges that monitoring surveys not be initiated until such commitments are in place.

Selecting Target Samples

Earlier work summarized in the committee's Phase I report showed that, among many young people, enlistment propensity tends to firm up by or before the end of high school. Although students provide answers to propensity questions as early as 8th grade in the Monitoring the Future (MTF) surveys, the proportions providing "definite" answers are much higher among 12th grade students. Senior year propensity measures, however, can be complicated by the likelihood that many seniors who eventually enlist make their decisions and commitments before leaving high school; accordingly, their "propensity" answers might better be described as reports of decisions already taken. This is clearly consistent with the often reported finding (Fishbein and Ajzen, 1975; Ajzen and Fishbein, 1980; Albarracin, Johnson, Fishbein, and Muellerleile, 2001) that the shorter the time interval between the assessment of intention (or propensity) and the observation of behavior, the better the prediction (i.e., the higher the correlation).[1]

Taking into account these factors, the committee recommends that the lower age boundary for target samples be about age 16-17. If there is an interest in propensity or related factors at lower ages, they could be the subject of limited special studies rather than monitoring surveys. By age 16-17 young men and women have had to confront questions about their next steps after high school (college, military service, civilian employment), so this seems a good lower age bound for target samples. Specifically, for reasons spelled out below, we recommend the 11th grade of high school as an optimal start point for the target sample age range.

The end point for the target sample is less easy to specify; however, given the increased recruiting attention to college students, the committee recommends that inclusion extend to at least age 23—the point by which most young adults either have completed college or are relatively unlikely to do so. We leave open the possibility that it may prove useful to extend the age span a bit farther; however, we think it unlikely that it would be cost-effective to extend it much beyond age 25.

Use of Self-Completed Questionnaires

There are very large potential cost advantages in surveys employing self-completed questionnaires. There are also constraints and limitations. The committee considered several important constraints and judged them acceptable for the proposed monitoring surveys.

[1]The fact that YATS excluded respondents who had already committed to the military, while Monitoring the Future included all high school seniors, helps to explain why, although both surveys found strong evidence that propensity does predict enlistment, MTF found stronger intention-behavior relationships than did YATS (see, e.g., Bachman, Segal, Freedman-Doan, and O'Malley, 1998).

Self-completed questionnaires, particularly when carried out on a large scale, tend to be limited to closed-ended multiple-choice question items. In spite of these restrictions in question-and-answer format, self-completed questionnaires have been used successfully in monitoring surveys (e.g., the Youth Risk Behavior Survey, Monitoring the Future). In contrast, YATS used telephone personal interview methods, which allow for open-ended questions and for potentially complex question branching strategies. Interestingly, even in the YATS telephone interviews, the survey data that tended to be most useful for monitoring purposes involved closed-ended items. Accordingly, the committee would find it acceptable to use self-completed questionnaires for monitoring purposes, even though that may constrain the survey questions to closed-ended items. (Open-ended items can be included in self-completed surveys, although they add costs, complexities, and time delays.) The committee recommends that other approaches (focus groups, smaller scale interview and elicitation studies, etc.) be used to develop and evaluate the closed-ended items.

Another extremely important limitation of self-completed questionnaires is that the target respondents must be able to read and understand the questions. Limiting the target population to literate respondents could be a serious bias in many studies, and even in the surveys proposed here this constraint needs to be kept in mind. Nevertheless, given a primary focus on potential military recruits, and given current and future military requirements for literacy among all Service members, the committee considers the literacy constraint tolerable.

One other dimension of self-completed questionnaires to be noted involves respondent motivation. All survey methods require some degree of motivation on the part of respondents, but arguably there are some methods that are more inherently motivating than others. Face-to-face personal interviews can be particularly motivating, because some degree of personal relationship is established between interviewer and interviewee. Telephone interviews may develop such a relationship to some extent; however, in a period in which telemarketing has imposed greatly on the good will of those still willing to answer their phones in person (rather than use an answering machine), the motivational value of phone interviewing may be debatable. In any case, one of the factors to be considered in developing any survey strategy is respondent motivation, and that will be particularly important for the surveys proposed here. The committee thinks that topics related to career choice are of great interest and importance to most of those in the target survey population and that, with sufficient care, self-administered questionnaires can be developed that will maintain respondent motivation.

Use of School-Based Administrations

Why survey in schools? To paraphrase Willie Sutton's bank robber comment: because that's where the students are. School-based survey administrations are wonderfully cost-effective—at least from the standpoint of survey researchers. Teachers and administrators, however, are becoming increasingly dubious about school-based surveys, partly because they seem to be proliferating, and also because the increased demands for student performance and school accountability have made classroom time a more precious commodity. The result is that obtaining cooperation for school-based surveys is increasingly difficult.

Nevertheless, in spite of the difficulties, there is much to recommend school-based group-administered self-completed questionnaire surveys: the sampling is fairly straightforward and can be quite accurately representative; response rates can be quite high among students; and, once again, it can be very cost-effective. Moreover, school-based surveys can also be treated as a baseline or starting point for panel surveys (repeated surveys of the same respondents over time). Panel surveys involve some respondent attrition, to be sure, but the base-year data can provide considerable information about the lost respondents (and how to make statistical adjustments to compensate for the loss).

Matters of Entry, Motivation, and Acceptability

A Pentagon-sponsored survey focused on military propensity, asking only questions about military jobs and missions, is likely to be acceptable to some school administrators and less so to others. Moreover, in most schools there is likely to be a range of parent and student reactions to a "military survey." Pentagon sponsorship should, of course, be acknowledged; but if the survey were a joint undertaking sponsored by the DoD and others, and if the content reflected that joint sponsorship, the survey might be more attractive to all concerned. Moreover, in order to do a good job, as well as to be broadly acceptable to school personnel, parents, and students, the military portion of the questionnaire content should be balanced in its items about military service, military working conditions, duty to country, and the like. If the survey comes across as promilitary rather than balanced in its approach, it is likely to generate considerable controversy and resistance in some communities and school districts. An emphasis on national service might aid in making the survey more broadly attractive.

A closely related point is that if personal identification of respondents is requested in order to permit follow-up surveys of some respondents, great care must be taken to ensure, and communicate, that no identifying information or individual survey responses will be available to anyone other than those conducting the survey. In particular, it must be clear to everyone that survey responses and the names of participants will not be made available to recruiters. This is not the sort of thing that can be mentioned once with the expectation that everyone will take note and remember. Rather, it is something that will need to be repeated often, to a number of relevant audiences.

Proposed Features of a Monitoring Survey Series

The committee considered a number of survey designs and options in its deliberations. None of the options was ruled out; to the contrary, it was considered important that multiple methods be employed at various points. In particular, the use of focus groups and individual personal interviews with open-ended questions was recommended as means of forming and improving questionnaire items. Nevertheless, consistent with the discussion of broad issues outlined above, the committee developed a design that it recommends for a monitoring survey series. We outline some key features below.

A Cohort-Sequential Design Using Self-Completed Questionnaires

The monitoring design proposed here would obtain new and relatively large samples of high school students annually and then track subsets of those students on a regular schedule during subsequent years. This is a cost-effective research strategy for generating descriptive data on youth and young adults, and it can permit disentangling changes over time that reflect age differences, cohort differences, and secular (historic) trends. (See, e.g., Johnston, O'Malley, and Bachman, 2003a, 2003b, 2003c, for illustrations of all three kinds of changes in substance use.) The design has a bonus feature: because it tracks the same individuals across time, it can be the basis for panel analyses exploring individual changes and growth trajectories, and it can distinguish chronological (and possibly causal) ordering of events.

Base-Year Surveys of High School Students

The committee recommends that the starting point in the cohort-sequential design consist of school-based group-administered self-completed questionnaire surveys of 11th grade students throughout (and representative of) the United States. As noted earlier, 11th grade is a point by which many students have confronted choices about their next steps after high school but before very many have made firm commitments to further education or to military service.

Most students in 11th grade are age 16 or 17, thus the choice of 11th grade as the starting point sets a fairly narrow age boundary. The choice of 11th grade rather than a particular age is governed, in part, by practical considerations involved in school-based surveys; however, the choice of a grade-based rather than age-based starting point also takes account of the fact that decisions about next career or education steps tend to be firmed up during 11th grade. In any case, we judge that grouping students by grade rather than by age is well suited to the purposes of the research.

For the reasons outlined above, the committee recommends 11th grade as the earliest point for the school-based monitoring surveys. Although surveys at earlier grades (and ages) could reveal things about the developmental progression of propensity, such could be the subject of special studies rather than an ongoing monitoring effort. The committee did not, however, reach a firm conclusion about whether 11th grade was the only grade worth including in school-based surveys. An obvious alternative would be to include both 11th and 12th grades, and we can see advantages as well as disadvantages of such a strategy. Later in this chapter we outline in greater detail two alternative strategies and consider some of the costs and benefits of each. One strategy involves sampling only 11th grade students and then tracking them with follow-up surveys, perhaps as often as once a year. The other strategy involves sampling both 11th and 12th grade students, with follow-up surveys every two years. It is important to note that either strategy will take time to develop the full range of respondents.

Periodic Tracking of Subsamples

Monitoring the Future tracks its panel respondents by mail every two years, with half of them resurveyed 1, 3, 5, 7, 9, and 11 years after high school and the other half resurveyed 2, 4, 6, 8, 10, and 12 years after high school. (Later follow-ups take place at modal ages 35, 40, and 45.) This every-other-year schedule works reasonably well for monitoring shifts in substance use, as well as for most other purposes; however, the committee notes that follow-up surveys conducted on an annual basis would provide greater detail in tracking changes in propensity, time of enlistment, and other related factors.

Anything more frequent than annual surveys would probably be burdensome to many respondents, and in any case seems unnecessary for monitoring purposes. Mailed self-completed questionnaires, accompanied by a modest "honorarium" check and coupled with phone follow-up prompts when necessary, can produce reasonably high response rates (even among those on active military duty). Some panel attrition occurs;[2] however, taking account of the initial base-year information obtained from all respondents often permits satisfactory statistical adjustments.[3]

Scheduling of Surveys Throughout the Year

The committee sees advantages in spreading surveys throughout the year in order to monitor more sensitively those military-related events that occur at unpredictable times during the year. As examples, surveys spread across the year might have provided some indications of short-term (versus longer term) impacts of events such as the World Trade Center attacks on September 11, 2001, and the initiation of the Iraq war in March 2003. Survey data spread across the years could be pooled across one-year intervals, or longer or shorter intervals, for various analysis purposes. Also, some analyses might need to take account of seasonal fluctuations in key indicators. It should be noted that some cost saving could result from spreading data collection across the year, particularly with respect to the follow-up surveys, simply because there would be fewer peaks and valleys in staff activities involved in carrying out the surveys.

Although spreading surveys throughout the year is judged to be desirable and quite attainable with the follow-up surveys, that objective cannot be met with high school surveys. In-school surveys cannot be carried out during summer vacation, nor are the first and last few weeks of school likely to be acceptable for surveys. Spreading the in-school surveys across about a six-month interval (October or November through March or April) is probably most feasible. Another approach would be to concentrate in-school data collections (to the extent possible) in two periods approximately six months apart (e.g., October and April). Complex sampling decisions would be needed in order to determine whether

[2]For example, MTF mail follow-up surveys one or two years after the 12th grade in-school survey have yielded returns of about 65 percent in recent years. Longer term response rates in that study (ages 30-40) have been 50 percent or higher. Response rates to mail follow-up surveys can vary markedly, depending on a number of survey parameters. These include length and interest level for the questionnaire, incentives (i.e., payments) used, and extent of tracking efforts to persuade delinquent respondents to return questionnaires. Thus it is not possible to predict in advance what response rates would be possible in the absence of knowing a great deal about such parameters. The MTF experiences noted here are merely suggestive; a DoD survey could perhaps do better, and it certainly also could do worse.

[3]As one example, Bachman, Freedman-Doan, Segal, and O'Malley (2000) made use of imputation techniques to adjust for the fact that panel attrition was greater than average among young men with high military propensity. Another approach is poststratification, in which differential weights are used to "reconstitute," in effect, the initial distribution on important variables.

schools could be offered much choice about when during the year they would be surveyed. Obviously, the greater the choice permitted, the greater the likelihood of obtaining a school's participation. It might prove possible to develop adjustments in seasonal data (taking account of characteristics of schools that fall into each of the intervals being analyzed) so that flexibility in school survey scheduling could be tolerated while still taking advantage of some spread of survey administration times throughout the school year.

Follow-up surveys can be scheduled throughout the year, and that scheduling could (and probably should) be independent of (orthogonal to) the timing of an individual respondent's base-year survey date. Thus, for purposes of exploring short-term impacts of various events during the year, the follow-up surveys spaced evenly throughout the year would be the better source of data.

Sampling Strategies for In-School Samples

A multistage stratified random sampling approach is proposed in which the first stage is the geographical area, the second stage is the school, and the third stage is the student. Schools are selected with probability proportionate to (estimated or reported) size. In smaller schools, all students in the relevant grade are targeted; in larger schools, a sample of students is selected (by sampling classrooms, not individual students). These procedures have worked reasonably well in Monitoring the Future across several decades; however, various other approaches are available for drawing school-based samples. Such sampling details can be developed by the relevant contractor, should the decision be made to conduct such surveys.

Sample-related issues to be kept in mind are that school refusal is a source of potentially important biases in survey results, as are respondent refusals; accordingly, efforts should be made to avoid such losses. It is also important to keep in mind that once a school is recruited and administration arrangements made, the marginal costs of additional students included in the sample are very low; consequently, it is not at all costly to have a fairly large number of respondents in school-based group-administered surveys. The primary factor influencing costs is the number of schools; that is also a key factor in determining how accurately the survey represents the target population. In Monitoring the Future, surveys of larger than 100 schools have always been judged necessary in order to have adequate levels of accuracy (the range for samples of 12th grade students has been from 124 to 146 schools). Other school-based surveys, such as those conducted by the Department of Education, have often used larger numbers of schools (with correspondingly higher costs).

Once the sample of schools has been obtained, it is relatively cost-effective to sample at least 100 students per school.

The committee recommends samples of 10,000-15,000 students per year (per grade). Among the key advantages of large sample sizes is the improved ability to provide estimates for relatively rare subgroups; race-ethnic subgroups are an obvious example, but another example is students with high military propensity—particularly women with high propensity Another advantage of large sample sizes is that it permits the use of multiple questionnaire forms; such an approach would include key "core" information (including demographics, propensity, etc.) in all forms but could cover a wider range of topics (or a wider variety of questions about the same topics) without appearing repetitious to respondents.

Additional Options for School Surveys

Once a school has been recruited to participate in the data collection, additional opportunities arise for collecting relevant data. This temptation must be resisted, or at least considered carefully, because the risk is that attempting to exploit the additional possibilities may have the very undesirable consequence of causing the school to withdraw (or refuse to grant in the first place) its agreement to participate in the survey. With that caution clearly in mind, we note that participating schools offer an opportunity to survey counselors. Although collecting data about specific students would almost surely be inappropriate, counselor data could be used to characterize the school as a whole, including the school atmosphere as it pertains to military service, recruitment, etc. Counselors could be surveyed about their own views, but also about their perceptions of school climate and perhaps certain relevant school practices. Counselor surveys might be Internet-based, if that seemed preferable. A modest payment (honorarium) for participation, as well as clear specification that participation is voluntary, would go a long way toward securing school agreement for such an additional option, as well as ensuring high levels of counselor participation.

Sampling Strategies for Follow-up Monitoring Surveys

The surveys of students should include respondents' names and mailing addresses (on separate computer-matched numbered pages), and respondents must be persuaded of the security of this information and thus be willing to provide it.[4] The listing of names and addresses, along

---

[4]A valuable additional piece of information would be Social Security number, because this could be used for military record checks (to include enlistment, separation, and per-

with the demographic and other data that can be matched with the names, becomes an exceedingly valuable base from which to develop target follow-up samples.

The follow-up sampling strategy is different from the base-year strategy in a number of respects—most notably because the marginal cost of follow-up respondent surveys is far greater than the marginal cost of the base-year surveys. As a result, it is more important to be efficient in drawing the follow-up target samples. Because of the information available, such efficiencies are readily attainable.

The base-year surveys might involve fairly wide ranges of weights because of the variations in school size, numbers of students selected (which can be adjusted up or down to make administrations more convenient for the schools, thus facilitating school agreement to participate), and a variety of other (lesser) factors involved in sample selection. In follow-up samples, however, selection ordinarily should be inversely proportionate to base-year sample weights; the result would be an equally weighted and highly efficient follow-up target sample.

Specifically, *school-to-school differences in sampling weights should be taken into account and eliminated.* Thus, schools that happened to contribute relatively large numbers of respondents (each having correspondingly small base-year sample weights), would have relatively small proportions of their survey participants selected for follow-up compared with other schools. An important departure from equally weighted follow-up target samples is that *respondents of particular interest to the Services should be oversampled.*

In particular, respondents with high military propensity should be oversampled. These are relatively rare individuals, and that might be reason enough to overselect them; more to the point, they are particularly important to the Services and thus should constitute a sufficiently large proportion of the follow-up target samples to permit analyses. In contrast, individuals who expect to enter college should be undersampled, primarily because they represent such a large proportion of 11th grade students that it is unnecessary for analysis purposes to have them constitute an equally large proportion of the follow-up target sample. These and other adjustments in likelihood of inclusion in the follow-up target samples can be corrected through appropriate weights in analyses. Other possible adjustments include oversampling of minority group members

haps other useful data). However, requesting such information without a clear explanation is likely to alarm some respondents, as well as some parents and school officials. Even if an accurate explanation is provided (i.e., to permit a link to later military records, if any), that also would be likely to raise alarms. The committee recommends careful pilot testing before attempting to include Social Security numbers as part of identifying information.

or individuals with occupational or educational interest profiles that are of particular importance to the Services.

Additional Follow-Up Options

Because the follow-up monitoring samples discussed above would not exhaust the numbers of base-year respondents who provide names and addresses, a variety of other special studies could be undertaken using some of these individuals. One sort of special study could target some individuals for Internet-based follow-up surveys. If that approach worked for a substantial portion of those sampled, and if it proved cost-effective and otherwise equivalent or preferable to the mail follow-up approach, then it might be possible to collect a significant portion of follow-ups via the Internet. This approach could be explored carefully and then, if successful, gradually phased in as an alternative to mailed surveys. (Under present conditions, it could not entirely supplant mailed surveys without missing a significant subset of respondents who would not be able or motivated to use Internet-based survey approaches.)

Trade-Offs Among Various Cohort-Sequential Designs

A great many designs are possible within the parameters suggested above. For purposes of illustrating the issues involved, we contrast just two design possibilities. The simpler design, which we will call Plan A, involves annual surveys of 11th grade students with a subset selected for annual follow-ups for a period of six years. This would (eventually) generate data for the age range from 16-17 to 22-23, permitting analysts to make distinctions among differences interpretable as age effects, cohort differences, and period effects (i.e., secular trends or historic changes). It would also permit relatively fine-grained tracking of individual changes in attitudes, plans, and behaviors.

The second design, Plan B, involves annual surveys of both 11th and 12th grade students with subsets from each grade selected for three follow-ups on a two-year cycle. This design would track the 11th grade respondents from ages 16-17 to ages 22-23 (like Plan A) and would track the 12th grade respondents from ages 17-18 to ages 23-24. This design (again, like Plan A) would permit distinctions among age, period, and cohort effects, and it would permit tracking of individual changes (although the two-year intervals would be less fine-grained than the one-year intervals in Plan A).

Many variations are possible within each of these two designs. As one example, Plan A could be extended to seven years of follow-up so as to match the overall age span in Plan B. As another example, Plan A could involve a two-year cycle of follow-up (like Plan B), or Plan B could involve a one-year cycle (like Plan A). Moreover, additional designs could be explored; for example, in-school surveys could be conducted with both grades, but follow-ups (on either a one-year or two-year cycle) could be carried out using subsets from only the 11th grade samples (so as to have all panel data begin with propensity measured earlier than the senior year).

It is not our intention here to arrive at a final ideal design; such decisions can better be made when more is known about the overall funding commitments, the nature of sponsorship (i.e., DoD only or shared), the likely success of obtaining school participation, and a variety of other crucial details (in which, as we know, the devil resides). For present purposes, it is sufficient to use Plans A and B as aids in illustrating some of the many trade-offs that will have to be considered to arrive at a final set of design specifications. In the remainder of this section, we focus on several key issues that should be considered in making decisions about design; these include school recruitment, the logistics of survey administration, follow-up participation and response rates, numbers of cases and amounts of data per case, and—cross-cutting all the others—considerations of costs.

School recruitment. Recruiting schools to participate in school-based surveys involves considerable effort and costs. If schools were willing to have both 11th and 12th grade students surveyed, the 12th grade survey (i.e., using Plan B rather than Plan A) would be a relatively low-cost addition. One of the costs, however, would be that schools could be recruited to participate for only one year rather than multiple years. If a school were to participate for more than one year, then many students would be surveyed once in 11th grade and again, a year later, in 12th grade; such an approach would add to student burden and could complicate analyses in various ways. Another cost might involve lower school participation rates, if the greater burden of surveying two grades prompted some schools to decline. (We suspect such declines would be relatively infrequent, but it is a risk that must be taken into account.)

Survey administration. There are advantages to completing all survey administrations in any one school during a single day. Not only does that tend to hold down survey administration costs (we recommend that administration be handled by a professional survey staff rather than school personnel), but it also reduces opportunities for students to discuss the survey with each other and thereby potentially contaminate responses. The logistics of managing both an 11th grade survey and a 12th grade survey in the same school at the same time (Plan B) would be somewhat more demanding than an 11th grade survey only (Plan A); per-school costs would thus be higher for Plan B than for Plan A, but certainly much less than double. One consideration affecting logistics and costs is whether identical questionnaire forms would be used for the 11th and 12th grade surveys (in Plan B). We see little need for separate forms (given that a single question can be used to identify respondents' grade level, and such a question is a necessary control in any case). The cost and logistical advantages of limiting to one set of forms are evident.

Response rates. Plan B would produce a better sampling of young people age 17-18 (i.e., 12th graders) than would Plan A. Under Plan A, the data from those ages 17-18 would be obtained from mail follow-up surveys of those surveyed as 11th graders a year earlier; even with the best efforts at tracking target respondents and securing their participation, roughly one-quarter to one-third of those targeted are likely to be lost. In contrast, under Plan B data from those ages 17-18 would be obtained by in-school surveys of 12th graders; losses in student participation would be much smaller. (The MTF experience in recent years is that 12 percent in the 12th grade sample was lost, primarily due to absence; the resulting biases can be corrected, at least in part, by using special weights that take account of the rates of recent absenteeism reported by students who did participate in the survey.)

Follow-up participation rates. We noted above that substantial portions of target samples are lost in mail follow-up surveys. That would occur for the follow-ups in both Plan A and Plan B; however, there is room for subtle differences between the two plans. On one hand, the two-year intervals between surveys in Plan B would presumably result in lower respondent burden than the one-year intervals in Plan A, and that might contribute to higher response rates in Plan B. On the other hand, the two-year intervals permit more time for respondents to become lost to the researchers and perhaps also to lose interest, and that might contribute to lower response rates in Plan B. We are not sure how these countervailing tendencies would play out; they might simply cancel each other. In any case, we believe that with careful efforts to track panel respondents (including some form of "keep in touch" mailing no less frequently than once a year), coupled with some modest financial incentive and an effort to keep the questionnaire content interesting and not overly burdensome, the differences in follow-up participation between Plan A and Plan B are not likely to be large.

Numbers of cases versus amounts of data per case. Let us for the moment make the simple assumptions that Plan A would involve surveying 10,000

11th grade students each year and then follow a subset (say, a target sample of 3,000) of each such sample for five or six annual follow-up surveys, whereas Plan B would survey 10,000 11th grade students plus 10,000 12th grade students each year and then follow a subset (again, 3,000) of each for three biennial follow-up surveys. Under these assumptions, Plan A would yield roughly half as many total respondents as Plan B, but it would involve nearly twice as many data collections for each. Once fully under way, each plan would provide appreciable numbers of cases for any given year and any age within the range specified—that is, roughly age 16 to age 23; moreover, except for the 12th grade data, the numbers of cases at any particular age or year would be roughly equal. But this would be accomplished differently, with Plan B providing about twice the total number of individuals but Plan A providing more fine-grained detail for each.

For panel analyses, the two-year intervals would probably provide sufficient detail for most purposes. Moreover, key questions, such as those about entering and leaving military service, could be asked with specific dates (to the nearest month, etc.) and thus provide nearly as much useable detail with a biennial follow-up sequence as with an annual sequence. Accordingly, the committee judges that the loss of fine-grained detail in the panel data under Plan B is not a severe handicap.

The advantage of having twice as many total cases under Plan B than Plan A is, however, considerable. The opportunities for cross-validation, exploration of rare subgroups, and other analytic improvements are potentially quite valuable, given that sufficient resources are provided for extensive analyses of the data.

Cost considerations. We have mentioned some of the cost considerations above, but here we focus on them specifically. The first point to be made is that costs will be determined by a great many factors, most of which we cannot (and should not) try to specify here. What we can do is focus on a few of the trade-offs and make some broad observations about cost considerations. Perhaps the broadest observation to be made is that we cannot with confidence estimate which plan would cost more than the other. Clearly, as noted earlier, if 11th and 12th grade students were surveyed in the same schools at the same times, there would be considerable cost savings—quite possibly the Plan B school surveys would cost only half again what the Plan A surveys cost (i.e., the 12th grade survey might be added at half price).

One could consider that 12th grade in-school survey cost as being weighed against the cost of a mail follow-up of the 11th grade sample (in Plan A) one year later. It is hard to know which would be more expensive;

much would depend on those devilish details. Depending on how many follow-ups were used in each, the follow-up costs might be fairly similar under the two plans. If Plan A involved six annual follow-ups and Plan B involved three biennial follow-ups of approximately twice as many people (e.g., targeting 3,000 each from the 11th and 12th grade in-school survey samples), the total numbers of targeted follow-ups would be identical and the costs would be quite similar. The Plan B follow-ups would be a bit more expensive because of the extra mailings needed to keep track of respondents in the off years. It should be noted that in the illustration above there is an extra year of data under Plan B; specifically, the 12th graders followed up a third time would have modal ages of 23-24, whereas the 11th graders followed up a sixth time (under Plan A) would be age 22-23. If the third follow-up of 12th graders were eliminated in Plan B, the overall cost comparisons would be more similar; however, there would be less complete data available on college completion (and possible enlistment afterward) if only two biennial follow-ups were used for 12th graders.

SURVEY CONTENT

If a major purpose of the survey is to identify the determinants of propensity or to track changes in propensity and its determinants over time, then, at a minimum, it will be necessary to assess propensity, attitudes, perceived norms, self-efficacy, and the behavioral, normative, and control beliefs underlying attitudes, norms, and self-efficacy. These clearly should comprise the core questions in any survey design.

There are now fairly standardized instruments for assessing these variables (see e.g., Fishbein et al., 2001; Ajzen and Fishbein, 1980). However, as we shall see below, although it's possible to develop measures of propensity (or intention), attitude, perceived norms, and self-efficacy without going directly to the population of interest, one must go to that population to identify the salient behavioral, normative, and control beliefs underlying these variables. Thus, formative research is necessary prior to developing a fixed-item survey instrument designed to assess underlying beliefs.

Propensity, Attitudes, Perceived Norms, and Self-Efficacy

Propensity/Intention: Propensity (or intention) is typically measured by asking a person how likely it is that he or she will (or will not) engage in the behavior in question. For example, the respondent could be asked to indicate the extent to which it is likely or unlikely that:

I will join the military (Army/Navy/Air Force/Marine Corps) sometime in the next N (months, years)
extremely unlikely ___:___:___:___:___ extremely likely

Alternatively, they could be asked whether they "strongly agree," "agree," "neither agree nor disagree," "disagree," or "strongly disagree" with the statement.

Attitude: Attitude refers to a person's overall feeling of favorableness or unfavorableness toward performing the behavior in question. Although there are many ways to assess attitude, the most commonly used instrument to assess attitude is the semantic differential (Osgood, Suci, and Tannenbaum, 1975). For example, respondents could be asked to indicate whether:

My joining the military (Army/Navy/Air Force/Marine Corps) sometime in the next N months would be:

good	____:____:____:____:____	bad
wise	____:____:____:____:____	foolish
pleasant	____:____:____:____:____	unpleasant
enjoyable	____:____:____:____:____	unenjoyable

Perceived Injunctive Norm: The perceived injunctive (or subjective) norm refers to a person's belief that their important others think they should or should not engage in the behavior in question. Thus, for example, respondents could be asked to indicate the degree to which they agree or disagree (or think it's likely or unlikely) that:

Most people who are important to me think I should join the military (Army/Navy/Air Force/Marine Corps) sometime in the next N months.

_____ strongly agree
_____ agree
_____ neither agree nor disagree
_____ disagree
_____ strongly disagree

Perceived Descriptive Norm: The perceived descriptive norm is one's perception of what important others are actually doing vis-à-vis the behavior in question. Thus it would be important to ask questions such as:

How many people like you will join the military (Army/Navy/Air Force/Marine Corps) in the next N months?

_____ none
_____ very few
_____ some

_____ almost all
_____ all

or:

Out of 100 people like you, how many will join the military (Army/Navy/Air Force/Marine Corps) in the next N months?

and:

How many people do you personally know who have been in or are now in the military?

Self-Efficacy/Perceived Control: Self-efficacy and perceived behavioral control refer to one's belief that he or she has the necessary skills and abilities to perform the behavior in question, even under a number of difficult circumstances. That is, it refers to the perception that one could perform the behavior if one "really wanted to." Items to measure this either could use the semantic differential format. For example:

My joining the military (Army/Navy/Air Force/Marine Corps) in the next N months is:
up to me ____:____:____:____:____ not up to me
under my control ____:____:____:____:____ not under my control

or they could be put in terms of a certainty question, such as:

How certain are you that, if you really wanted to, you could join the military (Army/Navy/Air Force/Marine Corps) in the next N months?
certain I cannot ____:____:____:____:____ certain I can

Underlying Behavioral, Normative, and Control Beliefs

Generally speaking, there are also standardized items that can be used for assessing behavioral, normative, and control beliefs. For example, behavioral beliefs are usually measured with items using the following format.

My performing (Behavior X) will lead to/prevent (Outcome Y).
extremely unlikely ____:____:____:____:____ extremely likely

Injunctive normative beliefs are usually measured with items using the following format:

(Referent A) thinks I should perform (Behavior X).
extremely unlikely ____:____:____:____:____ extremely likely

Descriptive normative beliefs are usually measured with items like:

(Referent A) performed or is currently performing (Behavior X).
____yes ____no

Efficacy beliefs are usually measured with items using the following format:

How certain are you that, if you really wanted to, you could perform (Behavior X), even if (Barrier A) were present?
certain I cannot ____:____:____:____:____ certain I can

What the above illustrations should make clear is that, prior to developing a fixed-item instrument to assess the beliefs that underlie attitudes, norms, and self-efficacy, one must first conduct formative research to identify the outcomes, referents, and barriers that are salient for the population in question.

Thus one could go to a small sample of the population in question and ask the following open-ended questions:

To Identify Salient Outcomes:

- What do you see as the advantages of your joining the military (or a particular Service) in the next N months? That is, what are the good things that would happen if you joined the military in the next N months?

- What do you see as the disadvantages of your joining the military (or a particular Service) in the next N months? That is, what are the bad things that would happen if you joined the military in the next N months?

- What else comes to mind when you think about joining the military (or a particular Service) in the next N months?

To Identify Relevant Referents (and Important Others):

- Please list those people who would approve of your joining the military in the next N months.

- Please list those people who would disapprove of your joining the military in the next N months.

- Please list any other people you would talk to or whose opinions you would consider if you were trying to decide whether or not to join the military in the next N months.

To Identify Salient Barriers and Facilitators:

- Please list those things that would facilitate or make it easy for you to join the military in the next N months.

- Please list those things that would prevent or make it hard for you to join the military in the next N months.

Responses to the above questions can be content analyzed and the most frequently mentioned (i.e., the most salient) outcomes, referents, barriers, and facilitators can be identified. This information can then be used to develop fixed-item survey questions to assess underlying behavioral, normative, and control or self-efficacy beliefs. This type of formative research is also critical as a first step in developing advertising strategy.

Outcome Evaluation

In addition to assessing behavioral beliefs (or outcome expectancies), it is also necessary to assess the evaluation of the salient outcomes. Most often this is done as follows:

(Outcome A) is:
good ____:____:____:____:____ bad

However, with YATS, rather than evaluating outcomes, respondents were asked to indicate the extent to which a given outcome was important to them. Thus it's necessary to consider whether to assess importance, or value, or both.[5]

Similarly, in assessing behavioral beliefs, should one simply ask if joining the military (or specific Service) will lead to each of the outcomes, or should a YATS-type question be retained that asks whether a given outcome is more likely to be obtained from the military or from a civilian job? Moreover, if one takes this comparative approach, is the appropriate comparison military versus civilian or military versus college? Irrespective of the answer to this question, we would recommend that the core set of questions includes assessment of both propensity (or intentions) to continue one's education (e.g., go to college) and to join the civilian workforce (e.g., get a job).

Finally, YATS assessed beliefs about each Service as well as the military in general. Thus another question that must be addressed is whether

[5]Because there were very few changes in ratings of importance over time, and because, with few exceptions, judgments of importance were not related to propensity, we recommended dropping importance in our earlier letter report (National Research Council, 2000).

a survey should ask about the military or about each Service separately, or both? Similarly, should it include the reserve forces?

Distal Variables

Distal variables may be related to propensity, but theoretically they are assumed to exert their influence indirectly by influencing underlying behavioral, normative, and control beliefs. For example, gender differences in propensity should be explained by finding that men and women hold different behavioral, normative, or control beliefs about joining the military. Similarly, the finding that young adults from the South are more inclined to join the military than those from the North should be related to differences in the beliefs held by young adults in these two geographic areas. Thus an important question to consider is what distal variables should be assessed. First, it seems reasonable to consider demographic variables that are known to be related to enlistment. These would include such variables as age, gender, ethnicity, socioeconomic status, education, geographic location (region/urban-rural/state) and employment status.

Second, as pointed out earlier, particularly with respect to youth, a very important potential influence on propensity is the image (or prototype) that one has of the kind of person who pursues a given choice option (e.g., the image of the kind of person who enlists in the military). Not only is it important to assess the prototype per se, but of equal interest is the extent to which a person's own self-concept maps onto (i.e., is consistent or inconsistent with) the prototype. As a result, we recommend that assessments of both prototype and self-concept be included in the set of core items comprising the survey instrument. Specifically, participants could be asked to rate "a person who enlists in the military" as well as one's self on a series of semantic differential scales, such as: "wise/foolish," "aggressive/timid," "works well with others/works best alone," "strong/weak," etc. Formative research would be necessary to identify the scales most appropriate for assessing a military prototype. Finally, it may also be useful consider other distal variables, such as attitudes (e.g., toward the military, education, or war), as well as personality or individual difference variables, such as intelligence and sensation seeking (Zuckerman, 1979; Palmgreen et al., 1995)

The above considerations are focused on survey content. We have tried to identify the types of questions that are needed to identify and monitor changes in propensity and its determinants. The data obtained from these questions provide the kinds of information necessary for developing effective media campaigns as well as for evaluating the effectiveness of advertising (or other types of interventions) designed to increase propensity. However, as indicated above, there are many other issues that need to be addressed in designing survey research. For example, in order to evaluate the effectiveness of various advertising campaigns or changes in recruitment policy, it is necessary to obtain data at regular intervals over time.

We have recommended the use of a cohort-based sequential sample survey with a longitudinal component that in our view will provide the information necessary for both tracking propensity and its determinants and for developing more effective communications to increase and maintain the pool of youth with the propensity to join the military. Moreover, by following the cohort on a regular basis, it will be possible, at least in part, to evaluate the effects of current events, as well as changes in advertising and recruitment policies. At the same time, however, we recognize that other data need to be obtained in order to fully evaluate advertising effectiveness and other recruitment initiatives. For this reason, in our earlier letter report on YATS, we recommended that "a portfolio of surveys at different time intervals replace the current annual YATS administration" (National Research Council, 2000). Let us thus briefly consider two other possible surveys.

Influencer Surveys

As noted earlier, there is considerable evidence that the decision to join the military is strongly influenced by other people. While much of this influence should be assessed through measures of injunctive and descriptive norms, it is important to recognize that these normative beliefs are usually quite consistent with reality. One usually is quite accurate in one's beliefs about the normative proscriptions and behaviors of relevant others. Thus, it may sometimes be necessary to change the beliefs and attitudes of influencers. In order to do this, however, one must first understand why influencers (parents, teachers, friends, etc.) support or oppose joining the military, as well as why they are (or are not) inclined to recommend military service. Since this is a totally different target audience, a separate influencer survey is recommended.

It is important, however, to distinguish between two roles of an influencer. On one hand, an influencer can encourage/discourage or support/oppose enlistment (i.e., as injunctive normative influences—see e.g., National Research Council, 2003; Rutter, 1980). On the other hand, influencers may be viewed as transmitters or evaluators of advertising or other types of interventions. They may influence and shape the way a person evaluates a given advertisement, advertising campaign, or recruitment policy (Hornik, 1997). Considerations such as these suggest a separate survey to more precisely track exposure to, and to evaluate the effectiveness of, advertising campaigns or other policy changes.

Exposure to Media and/or Recruiters

Although there are many ways to monitor exposure, the committee recommends that, if at all possible, respondents be presented with current TV/print/radio/Internet ads (plus one or more dummy ads) and be asked whether they saw or read it and, if so, how many times? This can easily be done if the survey instrument is computer assisted or presented on the Internet.[6] In addition to these subjective estimates of exposure, we would also recommend that gross rating points, as well as other indices of time or space purchased, be tracked.

Ad evaluation: For each of the ads to which the respondent reports exposure, a series of questions such as the following could be asked:

Did you believe the ad? yes _____ no _____
Did the ad tell you something you didn't already know?
yes _____ no _____
Was the ad appropriate for you (or people like you)? yes _____ no _____
Was the ad interesting? yes _____ no _____
Did you like or dislike the ad? like _____ dislike _____

To assess indirect paths of effect: For each ad (whether or not the respondent reported exposure), a series of questions such as the following could be asked:

Which of the following people have you talked to about this ad?

Provide respondents with a list of influencers (based on open-ended surveys) as well as "none" and "other (please specify)" alternatives and ask them to check all that apply.

For each influencer checked:

Did (Referent A) like or dislike the ad?

To assess interpersonal contacts: In addition to asking whether the respondent discussed ads (or policies) with various influencers, it would also be important to know if they discussed joining the military with these others. Thus respondents' could also be asked such things as:

[6]As indicated above, because of changing technology, it is possible that follow-ups of the cohort sample could eventually be done using the Internet. If this were the case, tracking could also be accomplished as part of a cohort-based sequential sample survey design.

Who have you talked to about joining the military? Check all that apply.

Provide respondents with a list of influencers (based on open-ended surveys).

For each influencer checked:

Did Referent A support or oppose your joining? support ___ oppose ___

INTERACTION OF PURPOSE, CONTENT, AND DESIGN

In this chapter, we have recommended a survey design and a set of core questions to assess and monitor changes in propensity and its determinants. We have also recognized that influencer and advertising tracking surveys are also necessary to gain a more complete understanding of advertising effectiveness and the role that advertising, recruiters, and other influencers play in the recruitment process. We recommend that there should be at least two types of surveys: an annual or semiannual cohort-based sequential sample survey with a longitudinal component to monitor changes in, and provide an in-depth understanding of, the determinants of propensity. In addition, this survey should include a more general set of questions about exposure to media and interactions with recruiters and other influencers. In different years (or at different times), questions about more distal attitudes and values could be assessed.

While such an annual or semiannual survey should address most of the key questions concerning propensity and recruitment, assessments of whether one has been exposed to a current advertising campaign, whether one has talked to others about that campaign, or whether one has talked to a recruiter will clearly vary as a function of advertising expenditure and military policies concerning incentives as well as the number and placement of recruiters. Thus, we recommend a brief, continuous tracking survey to assess exposure to specific events, advertisements (or other recruitment policy changes), and the extent to which the respondent engaged in discussions with influencers about the campaign or about joining the military. This survey should also monitor respondents' evaluations of the ads they were exposed to, as well as provide data to allow one to track changes in propensity, attitudes, norms, and self-efficacy over time. We recommend that separate studies (or surveys) be conducted to address specific populations, such as influencers and younger populations.

4

Advertising Planning: Generative and Evaluative Approaches

Following the theory of core variables explained in Chapter 2 and the survey approaches presented in Chapter 3, this chapter reviews critical trends in *behavioral beliefs* and *outcome evaluations* that point to research strategies and methods to support planning and evaluation of advertising for military recruiting. As explained in Chapter 2, positive attitudes and intentions toward enlisting in the military are supported by perceptions that enlistment definitely leads to valued consequences or outcomes. Hence, it is useful to understand how the range of relevant outcomes and their associated behavioral beliefs relate to the decision to enlist in the military as a foundation for successful development and selection of advertising message strategies.

A key fact for the planning of advertising in support of military recruiting is that "in the late 1990s, the Services struggled to meet their recruiting goals and in some cases fell short" (National Research Council, 2003, p. 1). This period of difficulty in meeting recruiting goals arose from a combination of factors. A strong national economy during the 1990s gave youth access to plentiful employment alternatives. At the same time, there was a dramatic increase in college enrollment and a continuing decline in youth interest in military service (National Research Council, 2003, p. 256 and 271). However, in addition to the availability of civilian job opportunities and youth interests in higher education, the increasing difficulty in meeting recruiting goals in the late 1990s can also be viewed in light of the direction of youth attitudes toward military service and corresponding trends in youth *behavioral beliefs* and *outcome evaluations* related to military service.

YOUTH BELIEFS AS A BASE FOR ADVERTISING PLANNING

Overall youth attitudes toward enlistment have often been studied with structured survey questions designed to track year-to-year changes in the propensity to enlist. Figure 4-1 shows that propensity to enlist among high school males declined between the mid-1980s and 2001. The proportion saying "definitely will" enlist declined from 12 to 8 percent between 1980 and 2001. At the same time, the percent saying "probably will not" or "definitely will not" increased from about 40 to about 60 percent. Figure 4-2 shows a similar trend for women. The downward trend in the highest propensity group coupled with the accumulation of a very substantial segment in the most negative group (the group at the top of Figures 4-1 and 4-2) indicates the importance of discovering whether changes in youth perceptions of specific outcomes and belief expectancies

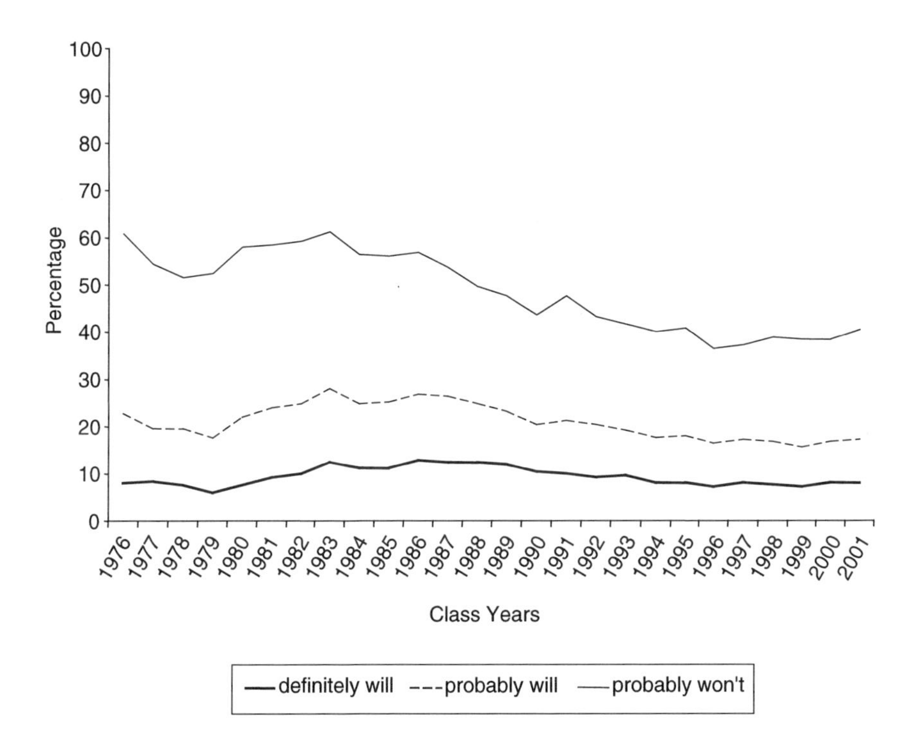

FIGURE 4-1 Trends in high school seniors' propensity to enter the military: males, 1976-2001.
NOTE: The spaces between the lines show the percentages in each of the three propensity categories.
SOURCE: Data from Monitoring the Future surveys.

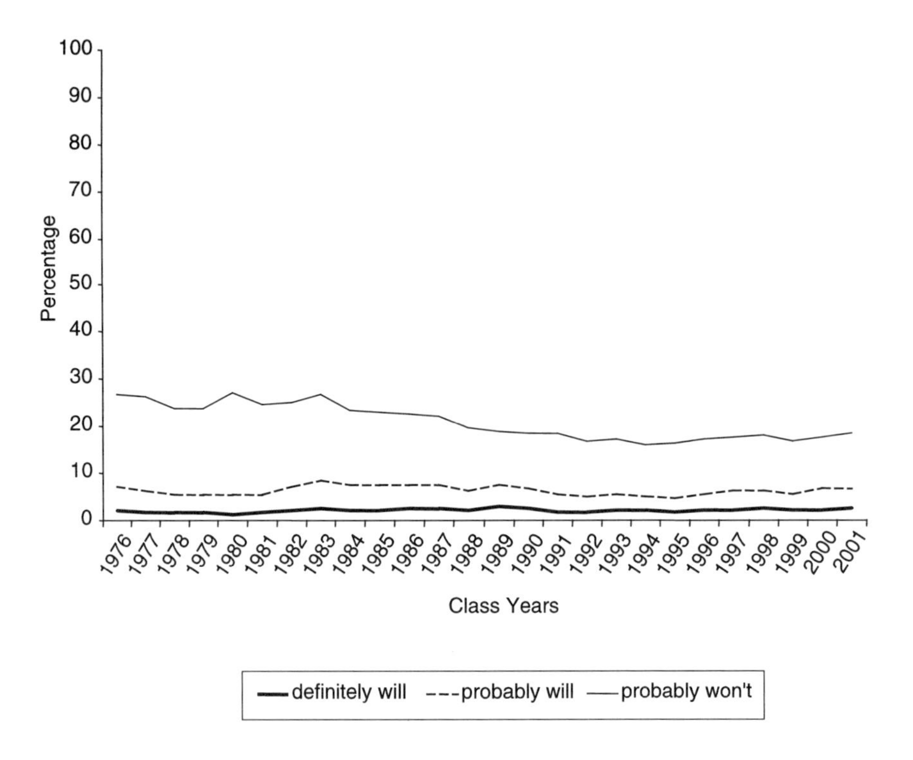

FIGURE 4-2 Trends in high school seniors' propensity to enter the military: females, 1976-2001.
NOTE: The spaces between the lines show the percentages in each of the three propensity categories.
SOURCE: Data from Monitoring the Future surveys.

offer potential explanations. Such findings could give rise to message strategies specifically addressed to the relevant outcomes and behavioral beliefs.

The potential for specific youth outcome evaluations and behavioral beliefs to relate to the continuing decline in the propensity to enlist was shown in a Defense Manpower Data Center report contrasting year-to-year results from the Youth Attitude Tracking Studies (YATS) (Lehnus, Srokowski, and Daniels, 2000). For example, between 1992 and 1999 the percentage of males ages 16 to 24 who evaluated the outcome of duty to country as "extremely important" or "very important" dropped from about 70 to about 57 percent (Lehnus et al., 2000, Figure 12A). During the same time period, the percentage of males ages 16 to 24 reporting the behavioral belief that duty to country could be more likely to be fulfilled in the military also declined from about 43 to 28 percent, while the percentage saying that this outcome could be better fulfilled in a civilian context increased from about 7 to 33 percent (Lehnus et al., 2000, Figure 12B). The 1992 to 1999 YATS surveys also showed a similar decline in the importance of duty to country among women ages 16 to 24. Women were also more likely than men to associate the goal of duty to country with civilian rather than military work.

The 1992 to 1999 time period was one of increasing economic prosperity, and corresponding changes in other behavioral beliefs concerning military versus civilian attribution are also informative. For example, for males the percentage saying that the outcome of working in teams is "extremely important" or "very important" increased consistently from about 71 to 77 percent. However, the behavioral belief that working in teams is an outcome to be realized in the military decreased from about 32 to 21 percent, while the civilian attribution increased from about 7 to 17 percent (Lehnus et al., 2000, Figures 10A and 10B). At the same time, military attribution of the outcome of preparation for a future career or job, a long-time positioning concept for military service, declined from about 20 to 13 percent for the military, while civilian attribution increased from about 17 to 28 percent (Lehnus et al., 2000, Figures 14A and 14B).

Since 1999, youth interest in military service has taken shape in the context of major national and international events as well as a dramatic deterioration of economic conditions involving major business firms, entire industries, and the overall availability of employment opportunities in the economy. Possible effects of these changing circumstances can be examined by comparing the results from the 1999 YATS survey with the Department of Defense (DoD) Youth Polls conducted in 2001 and 2002.[1] These are surveys of youth who have never served in the military and who were neither accepted for military service at the time of the survey interviews nor enrolled in postsecondary reserve officer's training corps programs.

A comparison of the results for these surveys shows that the propensity to enlist increased immediately following the events of September 11, 2001, for youth ages 16 to 21. The DoD Youth Poll of October-November

[1]The sampling frame for the YATS annual surveys was youth in the 16 to 24 age range, while the more recent DoD Youth Polls for 2001 and 2002 have sampled youth in the 15 to 21 age group. In this chapter, to provide for comparability of the survey results, the reported comparisons of these three surveys are based on a reanalysis of the data for the three surveys using only those respondents in the 16 to 21 age range. A further consideration is that sampling strategy as well as the time the survey is administered may be a differentiating factor. However, the consistency in response to the 24 attributes from one administration to the next suggests that the sampling strategies are similar across the surveys.

2001 found 15.5 percent of youth ages 16 to 21 saying they would "definitely" or "probably" be serving in the military in the next few years and 49.0 percent saying "definitely not." These results contrast with the 1999 YATS survey, which found 14.1 percent saying "definitely" or "probably" and 52.8 percent saying "definitely not."

The 2001 DoD Youth Poll also included questions concerning a range of 24 job-attribute-related outcome expectancies linking service in the military (or in a civilian job) to these attributes. A similar grouping of questions had been included in the annual surveys of the former YATS research program. Only two of the beliefs about outcome evaluations examined in the 2001 DoD Youth Poll showed significant and indeed substantial changes for both men and women: "doing something for your country" and "working as part of a team." For men, the proportion saying that doing something for your country is "extremely important" or "very important" increased 10.7 percentage points, from 57.1 percent in the 1999 YATS survey to 67.8 percent in the 2001 DoD Youth Poll. For women, the same comparison increased 15.5 percentage points, from 53.6 to 69.1. At the same time, the proportion of men saying that working as part of a team is "extremely important" or "very important" dropped 15 percentage points, from 75.6 percent in 1999 to 60.6 percent in 2001. For women, the same comparison dropped 16.4 percentage points, from 77.7 to 61.3 percent.

Turning to behavioral beliefs about the extent to which the 24 outcomes are seen as associated primarily with work in the military, the 2001 Youth Poll revealed significant changes in youth perceptions since the 1999 YATS survey with respect only to beliefs about "doing something for your country." For men, the percentage saying that "doing something for your country" was more likely to be realized through military service (than civilian work) increased from 25.7 to 41 percent. For women, the same comparison increased from 19.4 to 39.7 percent.

These dramatic changes with respect to reported youth beliefs concerning the importance of doing something for your country and working in teams reflect what youth ages 16 to 21 experienced and observed in the world around them during the years 2000 and 2001. What the future holds for these apparent changes in youth beliefs, and the possibility of changes with respect to other beliefs, will depend on the course of events. Clearly, the military action in Iraq during 2003 will be a significant factor in that regard. The most recent DoD Youth Poll (fielded in October-November 2002) found that responses to the standard question about the propensity to enlist had returned to the historical pattern established during the 1990s, with 14.6 percent of youth ages 16 to 21 saying they would "definitely" or "probably" be enlisting and 52.5 percent saying they will "definitely not" enlist (Sattar et al., 2002). The 2002 DoD Youth Poll did not also include questions concerning the 24 outcome evaluations and behavioral beliefs examined in 2001 and in the previous YATS surveys.

The belief changes revealed by the 2001 DoD Youth Poll findings, coupled with continuing national and international events of great significance, point to the need for continuous monitoring of the propensity to enlist with survey questions that also examine a range of youth behavioral beliefs and outcome evaluations relating to the decision to enlist in the military. The two major changes with respect to the outcome evaluations (the increase in importance of doing things for the country and the decline in the importance of working in teams) suggest the need to specifically examine the areas of duty to country and individualism as they relate to current youth interests. Indeed, as shown by the pre-1999 YATS surveys and the 2001 DoD Youth Poll, the direction of youth beliefs relating to doing something for your country appears to be closely associated with the direction of the propensity to enlist. An understanding of the current status of such beliefs is essential to effective development and planning of advertising message strategies. The findings also indicate the need to examine such considerations as the overall level of advertising to support military recruiting, the additional audiences (such as youth influencers) for recruiting related advertising, and the range of message strategies employed in the advertising support for military recruiting.

ADVERTISING PLANNING QUESTIONS

The dollar value of overall military advertising expenditures and recruiting bonuses has increased at least threefold since 1990 (National Research Council, 2003, p. 221; Office of Assistant Secretary of Defense [Force Management Policy], 2000). This recent and somewhat dramatic increase in military advertising expenditures was preceded by a period of substantial decline in the early 1990s, giving the recent increases the appearance of a recovery from an era of budget cutting. Figure 4-3 shows that advertising and recruiting bonuses declined by about 50 percent between 1990 and 1994.[2]

However, this apparent recovery in overall advertising support is compromised by a corresponding decline in the purchasing power of advertising media budgets. A recent report by the Rand Corporation has shown the purchasing power of the military advertising budget declined by about 60 percent between 1986 and 1997 (Dertouzos and Garber, 2003,

[2]This figure appears in the committee's Phase I report as Figure 8-1, with the addition of the original line showing the total budget. Note that the line for Joint Services advertising shows a planned 2003 budget increase that was later cancelled.

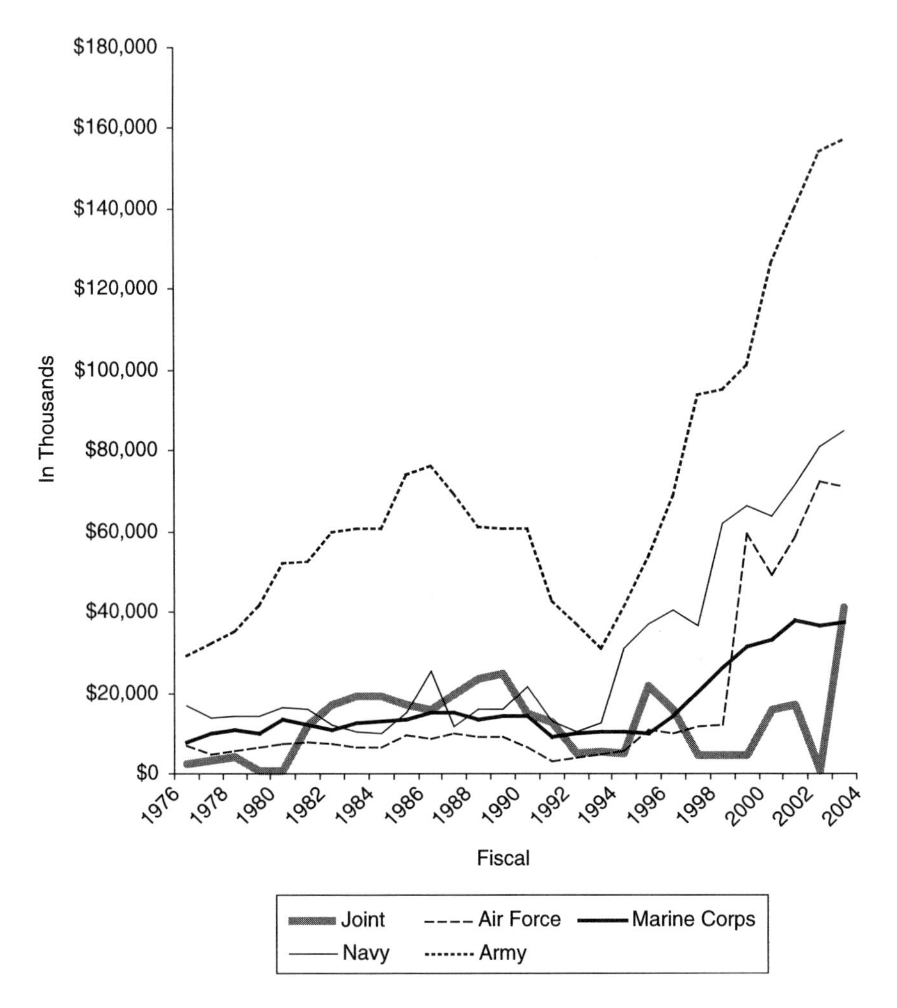

FIGURE 4-3 Enlisted advertising resources FY 1976-2003 in current dollars.
SOURCE: U.S. Department of Defense, 2002.

p. 7). This finding suggests that budget increases since 1994 may have served only to compensate for the long-term decline in the purchasing power of the year-to-year budget for military advertising. Furthermore, it seems likely that the declining marketplace presence associated with the decline in purchasing power of the advertising budget would have been exacerbated by what was an almost 50 percent reduction in the overall advertising budget between 1990 and 1994.

Importantly, advertising budgets are only one aspect of the deployment of advertising to support military recruiting. There are also critical decisions concerning the selection and definition of target audiences, the selection of the competitive frame (the choice context of competing options available to youth) underlying the advertising messages, and the specific strategic content of the advertising messages themselves. Indeed, in terms of sources of variance in advertising effectiveness in the marketplace, each of these considerations is just as important, and sometimes more important, than the amount allocated to purchase advertising space or time in the media.

The remainder of this chapter follows the general sequence of activities associated with the advertising planning process. The sequence begins with problem identification to support decisions concerning the competitive context, identification of the relevant audiences, development of a productive range of alternative message approaches from which to select the most promising approaches, and decisions concerning the amount of advertising.

Accordingly, the remainder of this chapter is organized as follows:

1. The competitive frame and audiences for military advertising.
2. Examination of audience member beliefs and goals.
3. Development of message strategies for military recruitment.
4. Allocation of resources to advertising message strategies.
5. Conclusions.

Youth Beliefs and Audience Definition

The competitive frame for military advertising can be viewed on two levels. The first-level competitive frame involves youth in the comparison of three broad areas of choice following completion of high school: (1) to pursue higher education, (2) to seek civilian employment, or (3) to seek military employment. This information-gathering and choice process involves comparison of three substantially different directions, each of which may include a variety of possible options. The second-level competitive frame involves the selection of a specific Service, and it applies to youth whose behavioral beliefs and outcome evaluations support the possibility of military service. This kind of more specific level of "brand choice" can involve already interested youth in the comparison or differentiation of the specific opportunities presented by essentially directly competing options from among the Services.

Military advertising message strategies have generally focused on the second-level competitive frame (or brand choice) that assumes the existence of some level of propensity to enlist. For example, the recent Army advertising claim that "there are 212 ways to be a soldier" differentiates the Army from the other Services in terms of the wide range of career or skill options available to youth, while the Marine Corps send their own differentiating attitudinal signal with their long-standing claim "The Few. The Proud. The Marines." However, the previously discussed trends shown by the YATS studies and the results of the recent DoD Youth Polls point to the value of focusing additional advertising message attention on the broad competitive frame in which youth view and contrast civilian employment, higher education, and military service.

The fact that over half of the youth ages 16 to 21 say "definitely not" with respect to the possibility of military service, and that this percentage continues to grow, indicates the need to develop an appropriate information campaign for the purpose of market development as opposed to the brand selection approach typically employed in military advertising. That is, the continuing low levels of propensity and high levels of negative propensity indicate the importance of exploring the role for advertising message strategies focused on the broad competitive frame of how youth of enlistment age contrast the options of civilian employment, continuing higher education, and military service. Beliefs relating to "doing something for your country" are particularly germane, and the recent increase in the outcome expectancy for this belief indicates that there is important potential for message strategies developed to specifically support youth beliefs in this area.

In addition to the role of advertising directed to the youth population, it is also important to examine the role of interpersonal communication and the extent to which it supports youth interest in military service. The 2002 DoD Youth Poll found that 53 percent of youth ages 15 to 21 said that the majority of their impressions about the military were based on information from friends and acquaintances, and 33 percent reported that a family member was the source of the majority of such information. When asked about whether they were influenced by these sources, 37 percent of the youth said the source had a positive effect on their likelihood of joining the military, 53 reported no effect, and 9 percent said the effect was negative.[3]

These results help demonstrate the role played by interpersonal communication sources (such as family members, friends, teachers, and others) whereby youth receive information and social support. The survey results suggest that the majority of these information sources are not

---

[3]Data analysis from October 2002, DoD Youth Polls of 2,000 youth ages 15 to 21 who have not enlisted and are not currently serving in the military (U.S. Department of Defense, 2002).

conveying a positive effect. This indicates there is a need to invest in information campaigns directed to audiences with whom youth communicate and consult when considering military service. Although there has been some recent development and testing of such advertising, the long-term practice has been to concentrate on advertising by individual Services designed to influence youth selection of those branches. While this approach may continue to play a leading role in the budgeting of military advertising, significant supplemental and continuing resources may be required to influence other critical and influential audiences in order to successfully address the continuing decline in the propensity to enlist and related trends in related youth behavioral beliefs and outcome evaluations.

The survey results from the former annual YATS surveys and the recent DoD Youth Polls have shown that significant recent events, such as on September 11, 2001, and recent war fighting actions can have important effects on youth beliefs and interest concerning service in the military. As previously discussed, youth interest in military service (as measured by the propensity to enlist) increased slightly and temporarily in 2001. A similar pattern in the propensity to enlist was observed during the time of the 1991 Persian Gulf war. Both the annual YATS survey program and the Monitoring the Future Survey showed the period of increased interest to be brief in the early 1990s, followed by a return to the ongoing trend of declining interest. These observed long-term patterns in propensity, coupled with what appears to be the possibility of rapid changes in related underlying outcome evaluations and behavioral beliefs, point to the need for ongoing surveys in the fashion set forth in Chapter 3. These observations also point to the possibility that these outcome evaluations are subject to reinforcement by specifically developed message strategies that could be conveyed in advertising directed to the youth population.

Behavioral Beliefs and Goals

The previous section identified two possible problems in the environment for military recruiting that might be successfully addressed by advertising. The first problem involves the downward trend in the propensity to enlist observed prior to 2001 and the possibility that the trend is driven to a significant degree by changes in the importance to youth of certain goals or outcomes and the attribution of these beliefs to military service (behavioral beliefs). The second problem involves the need for more supportive interpersonal communication from such groups as peers, parents, and other parties. These groups communicate with potential applicants and have the capacity to influence youth decisions concerning the decision to enlist in the military.

The range of behavioral outcomes examined in the former YATS survey program and the more recent DoD 2001 Youth Poll is one that developed in an ad hoc manner over a period of more than 20 years. There is a need for research to provide a more complete picture of the range of relevant beliefs held by the youth population. In particular, such research should include close examination of the entire range of beliefs relating to public service, duty to country, and the virtues associated with military service, such as personal sacrifice and concern for others. The perceived attractions of the possible outcomes or actions to be taken (such as enlisting in the military) are shaped by personal goals and value structures (Gutman, 1982). It is therefore important to more completely understand the entire structure of youth beliefs associated with public service, duty to country, and other virtues associated with military service. It can also be helpful to study the current language or word choice used by youth as they think and speak about these issues. Incorporation of language used by the audience facilitates the productivity of both the research process and provides helpful information for effective design of messages directed the audience of interest (see McQuarrie and Mick, 1999).

The questionnaires used in the 1999 YATS survey program and the 2001 DoD Youth Poll provide useful starting points for further research concerning youth values and beliefs. In addition to the standard measures of propensity, the YATS survey included questions about the importance of 26 behavioral outcomes and their related behavioral beliefs concerning the likelihood of successful pursuit of the values in civilian or military contexts. The 2001 DoD Youth Poll included 24 items based on the YATS approach.

Only one of the 26 items in the original YATS survey questionnaire examined the value of duty to country and the virtues associated with military service. Similarly, issues such as the importance of higher education and the importance of working in teams were examined by only one questionnaire item. For example, rather than pursuing the value of further education itself as a key outcome evaluation, the YATS questionnaire item focused only on the importance of obtaining money for education. Moreover, to manage the time burden of the surveys, the questionnaire items concerning outcome evaluations and their related behavioral beliefs were asked of randomly assigned subgroups of the total samples, thereby making it difficult to fully examine the relationships among the beliefs as well as their individual relationships with the measures of propensity to enlist.

To better understand the potential directions for advertising strategy, it would be helpful to conduct studies that more completely examine the range of beliefs related to the decision to enlist, particularly those associated with duty to country and youth interest in continuing with higher education. This is particularly challenging in that both of these areas of cognitive structure (beliefs related to the decision to enlist) ultimately bear on the very different competing choices available to youth (civilian employment, continuing pursuit of higher education, or military service). Choice situations involving divergent alternatives or seemingly non-comparable alternatives (such as military service and higher education) can be better understood by developing a more complete picture of the conceptual structure (beliefs) applied by the audience (Johnson, 1988).

The remainder of this section describes generative research approaches to reveal the belief concepts and language employed by youth and other research techniques that can be used to more completely describe and document conceptual structures for duty to country, interest in higher education, and related youth values relevant to the decision to enlist.

One approach is to conduct in-depth interviews with a relatively small sample of participants from the target audience. Study participants can be asked to contrast choice alternatives in their own words and to explain their thinking with increasing levels of probing. For example, one such generative technique is to employ pairwise or triadic comparisons of alternatives asking individual study participants to describe their perceptions of the similarities and differences of the alternatives presented to them. This approach is based on the repertory grid method, which organizes the concepts and language used by study participants to describe similarities and differences among groupings of stimuli (Kelly, 1955). The information provided by individual study participants can be contrasted and condensed as a way to identify the field of value or goal constructs used by study participants (Spiggle, 1994). The values or goals may vary in specificity from higher levels (more abstract), such as personal comfort, adventure, and prestige to lower levels (more product or choice specific), such as opportunity to learn job skills, compensation potential, and perceived difficulty of tasks (Howard and Sheth, 1969). In this case, the choices (or objects) for comparison involve civilian employment, higher education, and military service.

Once identified in generative research, the concepts and language used by the study participants can be used to develop a questionnaire for use in a field study of a representative survey sample of audience members. Following the approach used in YATS and the 2001 DoD Youth Poll, ratings for importance (or outcome evaluations) and attribution (behavioral beliefs) could be obtained for the various values or outcomes. Beliefs could be measured concerning the likelihood of each value or outcome being delivered or provided by each of a selection of alternative choices, such as civilian employment, higher education, and military service. Conventional analytical techniques such as factor analysis could be used to examine relationships among the values or outcomes, thus depicting the belief structure of the youth audience as it relates to the choices. The extent to which the values relate singly and in combination to such variables as propensity to enlist could also be examined.

The results of studies of this kind, including the approaches detailed in Chapter 3, would provide a more comprehensive and reliable view of the belief context (both outcome evaluations and behavioral beliefs) for youth decision making about military enlistment and would provide a basis for the development of alternative message strategies to inform youth concerning the possibility of military service. The results of such studies could also be used in the development of a revitalized annual tracking survey of youth values and propensity in the fashion of the approaches described in Chapter 3.

Message Strategies for Military Recruitment

Message strategies comprise four elements that identify (1) the audience of interest, (2) the desired response (such as switching one's brand choice or increasing the rate of use of a product category or specific brand), (3) the alternative actions or the competitive frame, and (4) the basic message argument (or promise) that may lead audience members to take the desired action (Overholser and Kline, 1975). Message arguments are also referred to as *concept statements* or *benefit statements.* Such a statement is written as a condensation of the core benefit so that effective comparisons can be made of the virtues of individual and specific alternative approaches to presenting and selling a product, service, idea, or organization.

For example, the following message argument "The Army gives you more choices" is one that may lead potential applicants to view the Army in a more favorable light in contrast to civilian alternatives or the other Services. All four elements of a message strategy are implied in that example: (1) an audience of youth making career choices, (2) strengthening of interest in or preferences for the Army, (3) a broad comparison to civilian or military alternatives, and (4) the idea that the Army has a greater selection of jobs from which to choose. The plausibility of such a message argument might also be viewed from the perspective of the Army, an organization that has a wide range of job descriptions to fill.

Another hypothetical message argument could be "The Army makes it easier to choose," which might speak to members of the youth audience with concerns about selection of a specific career direction. The role of such a statement is to provide a succinct promise that can guide the development of advertising messages. More effective advertising generally follows from message arguments that set forth an identity between a product strength and specific audience member beliefs. Such message arguments could apply to relevant youth outcome evaluations and behavioral beliefs and, at the same time, connect them directly to an appealing strength of military service.

A variety of message arguments can be developed for any product, service, or organization. The most useful arguments are generally those that bear most directly on values or beliefs that are associated with audience member decision making. Earlier in this chapter youth beliefs relating to the importance of duty to country and the association of this belief with military service were shown to be related to the propensity to enlist. Beliefs about public service and duty to country could be connected to military service with a message argument to the effect that the military, or a specific Service, is comprised of people of widely different job interests who are learning and fulfilling a wide range of assignments. Simply stated, such a message argument would be that "Military service offers the widest range of ways to serve one's country," thereby directly connecting a competitive advantage of the military (a wide range of work assignments providing opportunities for everyone) with a potentially important area of youth beliefs (duty to country). This is just one of many alternative message arguments that could connect the unique opportunities of military service. Many alternative message strategies could be developed and evaluated for their potential to reinforce relevant youth beliefs in such areas as duty to country, public service, career preparation, pride and accomplishment, and self worth, among others.

There are numerous approaches to testing the possible effectiveness of message strategies and message arguments. Moreover, it is important to adopt the policy of actively and continuously developing and evaluating alternative approaches that directly challenge existing message strategies and message arguments. The development and testing of challenge strategies can be incorporated in the ongoing work of research suppliers and advertising agencies. Armed with this information, organizations are better prepared to update existing message strategies, to appropriately change message strategies to respond to the competitive context, or to utilize a productive combination of mutually supporting message strategies to better inform relevant audiences.

To effectively select message strategies, communicators should develop a range of possible message strategies that are based on what is known about problems in the competitive context and an understanding of the relevant beliefs of the audience as they relate to youth choices in the competitive frame (or the available alternative actions, such as continuing in education, civilian employment, or military service). For military advertising, the message strategies of greatest interest will be those bearing most directly on the propensity to enlist and supportiveness of the audiences of friends, family members, and other influential people.

Message strategies for further consideration will be those that directly reflect what is known about the belief structure of the audience and the specific beliefs that are most relevant to propensity. The relationship between the belief structure and intentions to take the desired action (propensity to enlist in this case) provides important support for the selection of the message arguments with greatest potential.

It is also helpful to adopt the policy or practice of regularly (or continuously) challenging the existing communication strategies by developing and testing viable alternative message strategies. The potential attractiveness of alternative message arguments can be contrasted by presenting study participants (potential members of the intended audience) with simple written statements in what is called a "concept test" (Davis, 1997: Ch. 5). Or prototype examples of advertising messages (known as "comps" of print advertising and "animatics" for television commercials) can be tested in laboratory settings or by means of test-marketing selected market areas or media.

There are many approaches to testing prototype executions of possible communication strategies (Davis, 1997: Ch. 22). In-market tests can vary from small-scale use of print ads or television commercials to the use of selected market areas over extended periods to test message approaches and advertising budgeting levels. No single approach is ideal, and it is generally good practice to use multiple methods that effectively examine audience comprehension of the prototype advertising and the extent to which the advertising demonstrates the capacity to influence interest in military service.

In an experimental design context, this kind of research approach performs the role of a treatment check. The results can demonstrate whether the message argument, in the form that it is produced and tested, has the capacity to deliver the desired message to the intended audience. In this connection, it would also be important to adopt a consistent measure of propensity that reflects the approach most widely used in market surveys of propensity. Reliance on unstructured approaches based on group interviews with convenience samples (such as the commonly used focus group approach) is generally not an effective approach or method for purposes other than idea generation. Indeed, it is important that all research for advertising planning be placed in the context of a theory or model that focuses attention on the most relevant audience member beliefs in a valid and reliable manner. Such a theory was developed in Chapter 2 and is reflected in the analysis developed in this chapter.

Allocation of Resources to Advertising Message Strategies

The allocation of resources to advertising, whether to an organization's overall advertising budget or to specific communication objectives, products, or brands, is among the most important and challenging realms of organizational decision making. Well-chosen communication strategies, when carried out with effective message execution and media selection, can build information relationships with clients and organizational stakeholders that are necessary to year-to-year goal attainment.

Although scientific methods are regularly applied in support of advertising decision making, the area nevertheless remains an inexact science (see Dertouzos and Garber, 2003, for a recent review relating to military advertising). Presumably, all communication managers would like to make optimal decisions about the allocation of resources to advertising; however, the complexity of variables involved leads to the selection of generally "better" decision alternatives rather than proven identification of the "best" alternatives.[4] It is generally a year-to-year trial and adjustment process that includes a balancing of considerations of competitive communication activity, the direction of consumer interest, organization strategic objectives, and available resources. Historical perspectives on advertising spending, the availability of resources that might be allocated to advertising, and pragmatic judgment continue to play central roles in decision making about advertising budgets.

In this connection, one widely regarded approach to improving advertising decision making is to promote specificity in stating goals for advertising. Specific goals can focus decision making about the deployment of advertising resources and lead to the design of evaluation approaches that are more likely to provide for the possibility of valid and reliable results. This approach has been codified in the classic and still widely employed approach known as the DAGMAR (defining advertising results for measured advertising results) model (Colley, 1961, 1963). The premise of this model is that advertising is a communication process, and therefore advertising should be assigned specifically stated communication goals and then evaluated according to measured attainment of those stated goals. With respect to the planning of military advertising, such goals can be stated in terms of desired effects on measurable variables, such as specifically targeted outcome evaluations and behavioral beliefs as well as the propensity to enlist.

---

[4]For reviews of the scientific status of advertising allocation decisions see Ramond (1974) and Mantrala (2002).

It follows from this approach that military advertising should be allocated on the basis of the year-to-year measured information needs of the youth population (and other relevant audiences) and that evaluation of advertising performance in terms of the extent to which these information needs were met, as demonstrated by significant changes in such factors as audience awareness, beliefs relating to the decision to enlist, and intentions such as the propensity to enlist. Reliance on traditional economic analysis, with recruiting contracts as the dependent variable, would not be seen as necessarily productive in this regard because the communication effects of the advertising exposures would be embedded in a broad context subject to many other potential causes. In particular, the practice of keying the annual military advertising investment to year-to-year recruiting goal attainment, as measured by the number of enlistments, is to allow the process to be driven by such factors as the direction of the overall national economy, recruiter deployment decisions, and enlistment incentives and to ignore the ongoing information needs of youth as revealed by the continuous tracking of measures of youth beliefs relevant to the decision to enlist and measures of the propensity to enlist.

The current status of survey measures of youth beliefs and the propensity to enlist, discussed earlier in this chapter, provide an assessment of the orientation (or readiness) of the youth population to enlistment. As the events of the past 10 years have amply demonstrated, the Services can encounter difficulty meeting enlistment goals when key youth beliefs and the propensity to enlist decline to some recently observed levels. The reinforcement and maintenance of these beliefs and an agreed-on level of propensity should be central to advertising planning decisions, particularly decisions concerning the annual investment in advertising.

Earlier we posed specific questions about youth beliefs related to the propensity to enlist that could lead to the development of possible advertising message strategies. Once developed and tested at the message argument level, such concepts could be further evaluated in laboratory and in-market situations. The assessment could focus on the specific attainment of desired changes in such measures as changes in awareness of the desired strategic message elements, the levels of specified beliefs, and the propensity to enlist.

Three broad question areas can be summarized as follows:

1. Is there an erosion in youth beliefs or values that bear importantly on the propensity to enlist? If so, are there message strategies to effectively inform youth beliefs related to the propensity to enlist?

2. Is the overall budget level for military advertising sufficient, given the ongoing pattern of erosion in the propensity to enlist since 1980? The recent developments in the economy (declining civilian employment

opportunities) and in national and international events appear to result in temporary boosts to propensity. Although the favorability of the ongoing environment for military recruiting appears to be somewhat cyclical, the underlying downward trend in youth propensity to enlist will continue to periodically emerge as a barrier to reaching military enlistment goals.

3. Can messages strategies used for youth recruiting also sufficiently inform youth influencers, or are specialized message strategies needed to inform parents, teachers, counselors, community leaders, and other youth influencers?

The possible in-market potential of alternative message strategies could be tested in a variety of ways. Indeed, given the research design and measurement challenges, triangulation of the results of several evaluation approaches would be a good practice. These approaches could include studies conducted in laboratory settings, smaller scale in-market applications, and multimarket area test marketing of possible advertising campaigns. It would also be useful to revisit selected aspects of the former YATS survey approach that would provide for tracking of awareness, beliefs, and propensity both to diagnose potential communication issues and to track year-to-year effects of communication programs designed to influence awareness, beliefs, and the propensity to enlist.

Laboratory Testing

Copy testing techniques, such as provided by commercially available services or specially designed studies, can be applied in experimental design settings to contrast the potential effectiveness of alternative message strategies. Some of these approaches involve theater-style testing of mock-up commercials, others involve specially designed tests of portfolios of prototype print ads, or tests could be based on exposure to alternative strategic approaches produced on prototype web pages. In this approach, exposure to message strategies can be controlled and the capacities of the strategic directions to inform and influence various audiences can be evaluated and contrasted. For example, a conventional "2 by 2" factorial design could be employed to contrast the potential effects of two different message strategies on a sample of research participants selected to represent two audience groups of interest (a positive propensity group and a negative propensity group). Clearly, more complex designs could be developed depending on the alternative message issues to be contrasted and the types of audience groupings of interest. The value of employing an experimental design outlook is that such studies call for a research design to test specific alternatives, effective sampling techniques to represent desired audiences, and the use of valid and reliable measures of effects. Importantly, the exact features of the design should be based on a theory or model (such as the one developed in Chapter 2 and employed in this chapter) and be utilized to inform specific alternative choices relating to the selection of message strategies or other advertising planning questions. This is a more attractive approach for planning advertising than the qualitative research studies that are sometimes employed for advertising message evaluation, especially the highly subjective approach of focus group interviews.

Smaller Scale In-Market Experiments

Another approach is presented by the possibility of utilizing web sites in tandem with the selective use of conventional advertising media, such as newspapers, magazines, and television. Web sites (or controlled partitions within web sites) could be used to represent differing message arguments. Alternative message approaches could be presented on a random basis to web site visitors, or selected media vehicles (individual ad placements in newspapers, magazines, or television commercial placements) could be used to test the capacities of alternative message approaches conventional to interest youth in visiting specific web site locations. The web site technology would enable specific tracking to measure whether "information relationships" can be developed with individual youth and whether these relationships lead to applicant status, completed contracts, and successful completion of basic training and the first term.[5] In this approach, message arguments could be tested in terms of their capacities to attract interest and to guide audience members to further information available on web sites. The flexibility and specificity of media use in the approach would also allow for the testing of messages designed to reach specialized groups of youths or other audiences of interest. In this way, the message argument effects could be traced in terms of their capacities to influence individual youth to visit with military recruiters, to become successful applicants, and even to complete basic training.

In-Market Testing

In-market testing of alternative advertising messages, media selection plans, or budget levels is a challenging endeavor. There are difficulties in selecting (effectively matching) test market cities or regions, managing specific media vehicles to control spillover or contamination, controlling

---

[5]For a discussion of web sites and the concept of information as a relationship, see Eighmey (1997).

related factors such as recruiting efforts, and developing outcome measures that apply specifically to the objectives of the market test (Bogart, 1986; Dertouzos, 1989).

Nevertheless, test market approaches could be particularly useful with respect to the issues such the direction of specified youth beliefs and the continuing decline in the propensity to enlist. One approach to this problem is to invest in new and additional specific advertising directed to the youth population and to youth influencers. New campaign message approaches and new product introductions present opportunities for more effective use of test markets (Bogart, 1986, p. 368). Advertising to inform youth propensity and addressed to youth influencers would be new approaches in the marketplace and, in the fashion of a new product introduction, could present opportunities for test marketing and effective tracking of outcomes in test market areas.

This approach would be useful for deciding whether there is underspending on the overall level of the military advertising effort and whether specific campaigns directed to youth beliefs supporting propensity and to youth influencers could improve the productivity of the military recruiting process. Decision making with respect to these issues could be supported by in-market experiments or market tests that are of sufficient duration and that employ noticeable levels of advertising in the selected test areas. Survey methods could be used to track and compare awareness of the test advertising, key beliefs, and the propensity to enlist.

Annual Survey of Propensity and Related Youth Beliefs

As demonstrated by the recent changes in youth beliefs and propensity, there is a need for an annual survey research to update the previous tracking of propensity to enlist and to measure the wider range of related youth beliefs or values related to propensity. With respect to erosion of certain youth beliefs related to the propensity to enlist, we have identified the need to explore the potential impact of new message strategies concerning topics such as duty to country and other youth beliefs. An annual survey of propensity and related youth beliefs would assist in decision making concerning the need for additional advertising efforts to inform youth and the success of any such informative advertising campaigns in influencing the overall level of propensity and youth beliefs that might be the subject of such advertising.

CONCLUSIONS

The decade of the 1990s was a period of increasing challenge for military recruiting. A strong economy presented increasing work opportunities for civilian employment, youth interest in higher education increased, and certain youth beliefs became less associated with military service. These developments led to the point that annual recruiting goals were not met in 1999. Since that time, employment opportunities in the civilian workforce have become less plentiful, there have been highly visible events involving national security, and some new advertising approaches for military advertising have been introduced.

However, recent surveys of youth interest in the military have indicated a short-term increased interest in military service followed by what appears to be a return to the levels observed prior to September 11, 2001. Such a direction would be consistent with the pattern in the propensity to enlist shown by the YATS survey data for the period before and after the 1991 Persian Gulf war. These developments indicate the need to base advertising planning on a model of youth beliefs relating to the propensity to enlist and for continuous and consistent monitoring of those beliefs and the propensity to enlist.

Once the relevant youth beliefs (outcome evaluations and behavioral beliefs) are examined to an improved extent, it is recommended that alternative advertising message strategies be developed and tested on a regular basis to identify a range of possible message strategies beyond those that have been traditionally employed to support military recruiting. The possible effects of these message approaches should be tested using a theory-based approach and a variety of testing methods to assess the potential to use advertising to inform both the youth population and the audience of adult youth influencers. Some questions that might be addressed include:

- Will messages of duty, honor, and country result in a larger increase in propensity than other messages, such as those invoking personal challenge, camaraderie, training, and adventure?
- Will propensity increase more substantially if messages are directed at influencers rather than at young men and women?
- Over what range of expenditures is advertising cost-effective?

Finally, we recommend that military advertising be allocated and assessed on the basis of the year-to-year measured information needs of the youth population (and other relevant audiences) and that evaluation of advertising performance focus on the extent to which these information needs were met, as demonstrated by significant changes in such factors as audience awareness, beliefs relating to the decision to enlist, and intentions such as the propensity to enlist. This approach focuses on marketplace factors that can be directly influenced by advertising. The practice of keying the annual military advertising investment to year-to-year recruiting goal attainment, as measured by the number of enlistments, inherently focuses on such factors as the direction of the overall national economy, recruiter deployment decisions, and enlistment incentives and ignores the ongoing information needs of youth as revealed by the continuous tracking of measures of youth beliefs relevant to the decision to enlist and measures of the propensity to enlist. The variable nature (up and down year-to-year) of advertising budgeting encouraged by focusing primarily on enlistment goals seriously undercuts the capacity of military recruitment advertising to help maintain the year-to-year readiness of the youth population with respect to the possibility of military service.

5

Determining Optimal Levels of Advertising and Recruiting Resources

Generally, econometric approaches are the most suitable research designs for assessing the optimal levels of recruiting programs and resources. They can be applied to both natural and experimental data. Because they often can be applied successfully to natural data, they can save the often significant costs associated with a formal experiment. This can occur, however, only for resources and programs that have been implemented and for which there is variation cross-sectionally, over time, or preferably both.

Econometric methods can be used to isolate and identify the effects of existing resources, policies, and external factors affecting recruiting outcomes as well as their costs. There is by now a relatively well-developed body of econometric research that has identified some of the most important determinants of enlistment supply as well as the cost and effectiveness of various recruiting resources and the trade-offs among them. Estimates are based on the natural variation in key recruiting resources and outcomes (usually aggregated) that occur over time and across geographic locations.

In this chapter we first review current approaches as described in the recent literature and then suggest practical areas in which the research design can be improved (with best data sources) and that promise to produce the most reliable results. An implicit theme is that one cannot approach design issues in isolation. That is, to obtain better estimates of the effect of advertising, for example, improvements must be made in models that already account for other factors affecting supply in a solid way, and vice versa. Hence, it is probably a mistake to think in terms of the best model for estimating recruiter effects, the best model for estimating advertising effects, and so forth. In this multivariate framework of the actual recruiting market, one must adequately control for all factors to isolate the effect of a single factor.[1]

ECONOMETRIC APPROACH TO ENLISTMENT SUPPLY

From the beginning of the All-Volunteer Force, enlistment supply has been an ongoing topic of research.[2] Econometric studies of enlistment supply have used either aggregate national time-series data or panel data—that is, data over time disaggregated by some geographic level (e.g., state, county, Service-specific recruiting area). Early studies typically focused on highly qualified enlistments (H) and modeled H as a function of exogenous economic factors (X) and recruiting resources (R): $H = h(X, R)$.[3] These studies implicitly assumed that the supply of recruits with low qualification levels (L) was unlimited and that these recruits are costless to recruit.

Dertouzos (1985) and Polich, Dertouzos, and Press (1986) introduced the current generation of recruiting supply models. These models are distinguished by accounting formally for the role of recruiters' preferences, the recruiting technology, and recruiter incentives. Dertouzos (1985) argued that because it takes time and effort on the part of the recruiter to attract and process even walk-in recruits, a more appropriate formulation of the supply of highly qualified recruits adds L to the high-quality enlistment supply function: $H = f(X,R,L)$. L should have a negative effect on H in this formulation.

Recruits do not simply appear off of the street. Recruiters must seek them out and provide information about military service opportunities that may convince them to join. This activity requires recruiters to expend effort. The high-quality enlistment function is thus further modified to include recruiter effort: $H = f(X,R,L,E)$. Recruiter effort is unobservable to researchers. But Polich, Dertouzos, and Press assume that it depends on how high H and L are relative to the quotas that recruiters are given for these two qualification categories of recruits (Q_H and Q_L, respectively).

---

[1]This can be tempered somewhat by noting that a particular specification designed to measure advertising effects, for example, may control for other factors in a way that is not designed to produce structural estimates of their effects, but simply control for variation from that source. The point here is that this other variation must be accounted for.

[2]Nelson (1986) provides a useful survey of the voluminous studies conducted prior to the mid-1980s.

[3]High-quality enlistments are enlistments of high-school diploma graduates who score 50 or above on the Armed Forces Qualification Test.

When H (L) is low relative to Q_H (Q_L), recruiting is more difficult and recruiters must work harder. Therefore, $E = g(Q_H, Q_L)$. Substituting this expression into the function above for H gives $H = f(X,R,L,Q_H,Q_L)$. Most enlistment supply studies operationalize this general function by assuming that the natural logarithm of H, $log(H)$, is a log-linear function of the other variables:

$$log(H) = \eta_L logL + \eta_X logX + \eta_R logR + \eta_{QH} logQ_H + \eta_{QL} logQ_L + \varepsilon$$

The random error ε in this equation accounts for unobservable influences on $log(H)$. The coefficients in this equation are elasticities of H with respect to the variables in the equation. Elasticities show the percentage change in H due to a given percentage change in a given variable. For example, a recruiter elasticity of 0.4 indicates that a 10 percent increase in the recruiter force would lead to a 4 percent increase in H.

Warner, Simon, and Payne (2002) provide a detailed review of 15 econometric studies of enlistment supply conducted between 1985 and 1996. Some of the elasticity estimates from these studies are summarized below. Models are usually estimated with panel data—that is, data that vary by time (e.g., month, quarter, year) and cross-section unit (e.g., Service recruiting area, state).[4]

The log-linear model rationalizes the inclusion of quotas as factors affecting highly qualified enlistment. It implies that an increase in quotas will, by stimulating recruiter effort, increase enlistments. It is possible, however, that the effect of quotas depends on whether H is above or below Q_H and L is above or below Q_L. Daula and Smith (1985) and Berner and Daula (1993) pursue a different modeling strategy based on the concept of "switching" regression. Their approaches break the restriction of linear relationships between $log(H)$ and $log(Q_H)$ and $log(Q_L)$ and permit changes in quotas to have different effects depending on how high they are relative to H. Daula and Smith (1985) allowed $log(H)$ to switch between "supply-constrained" regimes $(H < Q_H)$ and "demand-constrained" regimes $(H > Q_H)$. Studying Army recruiting battalions, Berner and Daula (1993) allowed H to fall into three regimes: those that were highly supply-

[4] The proper estimation method with panel data depends on the form of the residual in the model. In panel data, the residual ε may be composed of three different factors. The first is a state effect, which is constant over time. The second is a time effect, which captures the influences of unobservable influences that are common to all states at a point in time. The third is an idiosyncratic factor, which varies randomly by state and time. Models with panel data can be estimated by using one-way or two-way fixed-effects models that control for unobservable state and time effects (Greene, 2002).

constrained, those that were producing around the quota, and those that were producing well above their quota.

Issues in Estimation of Advertising Effects

Two key issues in estimating the effects of advertising are (1) functional form, the shape of the relationship between expenditures or impressions and enlistments, and (2) dynamics, how advertising in one period affects recruiting in subsequent periods. The simplest form of relationship is a semilogarithmic relationship in which *log(H)* is related linearly to current or lagged expenditures or impressions. The semilogarithmic relationship imposes the restriction that each dollar increase in advertising gives the same *percentage change* in *H*. Another frequently assumed form of relationship is the log-log relationship, in which *log(H)* is related to the natural logarithms of current or lagged advertising. This form of relationship imposes the restriction that the elasticity of enlistment with respect to advertising is constant regardless of the level of advertising. A shortcoming of this specification is that the logarithmic transformation requires excluding those observations for which the advertising measure has a value of 0, often the case with military advertising.[5]

Dertouzos and Garber (2003) argue that the functional relationship needs to be flexible in order to estimate the effects of advertising over a wide range of levels of advertising. Furthermore, they argue that advertising must reach a minimum critical level before it has any impact on enlistment. Beyond this critical minimum level, increases in advertising increase highly qualified enlistments, first at an increasing rate and later at a decreasing rate. Finally, beyond some saturation level, advertising ceases to have any impact on highly qualified enlistments. A form that allows regions of both increasing and diminishing returns is the logistics function. Figure 5-1 illustrates logistics and linear relationships between the log of highly qualified enlistments and advertising expenditures.

The second important specification issue raised by Dertouzos and Garber is the timing of the relationship between advertising and enlistment. Advertising in a particular month is likely to affect highly qualified enlistments in future months, and the problem is how to specify the timing of the advertising-enlistment relationship. Some studies have imposed specific distributed lag relationships. One popular form of distributed lag relationship, the Koyck lag, imposes a geometrically declining relation-

[5]Alternatively, one can set the advertising measure equal to some small number. However, Hogan, Dali, Mackin, and Mackie (1996) reported that results were sensitive to this choice of number. For this reason, they entered advertising linearly in levels.

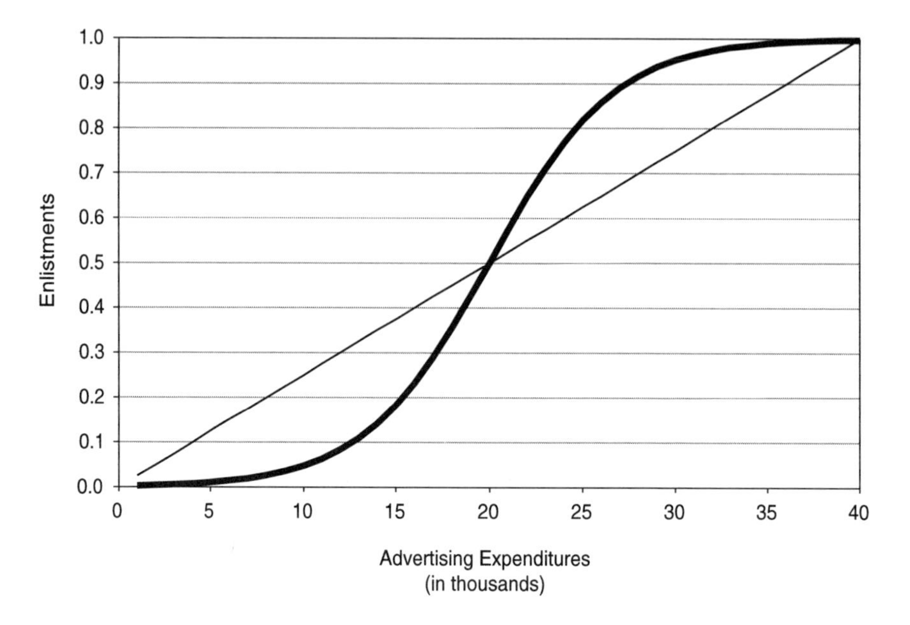

FIGURE 5-1 Hypothetical relationships between log of highly qualified enlistment and advertising: Logistics and linear cases.

ship between past advertising and current enlistments. That is, advertising has a larger near-term effect than far-term effect. Specific lag forms such as the Koyck are often assumed when time-series are short and there are insufficient data to accurately estimate a large number of lag parameters. Despite this virtue, they run the risk that the true relationship is not of the assumed form. With a long enough time series, it is possible to simply include a sufficient number of lags of advertising in the model and estimate the parameters by regression.

Estimates from Econometric Studies

Table 5-1 outlines the empirical strategies of 16 enlistment studies of male highly qualified recruits carried out between 1985 and 2001.[6] Eleven of these studies focused on a single Service. The factors that determine high-quality enlistment supply fall into three categories: (1) recruiting market factors (relative military pay, unemployment rate, youth popula-

[6]Nelson (1986) summarizes studies performed with data from the 1970s.

tion); (2) recruiting resources (number of recruiters, advertising budgets); and (3) recruiting policy variables (recruiting goals, enlistment bonuses, college benefits). Table 5-2 summarizes the findings from the studies in Table 5-1 with respect to recruiters and advertising.

Excluding Warner's (1990) negative estimate for the Air Force, estimates of the elasticity of highly qualified enlistments with respect to recruiters range from a low of 0.090 for the Air Force (Fernandez,1982) to 1.65 for the Army (Dertouzos, 1985). The Fernandez and Dertouzos estimates were obtained from data that spanned very short time periods. Estimates from longer time periods and periods over which recruiters exhibit more variation are more reliable. Furthermore, because the allocation of recruiters to geographic areas is likely to be correlated with unobservable factors that vary systematically across geographic areas, estimates based on models that include fixed geographic effects are probably less biased than are other estimates. In fact, studies with panel data employing fixed effects for geographic area or time (or both) have tended to yield smaller recruiter elasticity estimates.[7] The mean of the recruiter elasticity estimates in Table 5-2 is 0.55, implying that a 10 percent change in the recruiter stock changes enlistment of highly qualified recruits by about 5.5 percent.

As Table 5-2 shows, there are many fewer estimates of the effects of advertising than of recruiters. In the only experimental study of advertising effectiveness, Dertouzos (1989) reanalyzed the Ad Mix Test data. Unlike the original analysis by Carroll (1987), Dertouzos estimated positive but modest effects of advertising. However, flaws in the design of the test limited the usefulness of the data produced by that experiment. The table reveals the paucity of econometric estimates of advertising effects. A primary reason is the lack of data. The Navy, through a contract with PEP Research, Inc., was the only Service to systematically collect advertising data at any geographic level of detail throughout the 1980s.[8] Through a contract with the Department of Defense (DoD), PEP collected advertising data on all four Services by month by county for the period 1988-1997.

Warner (1991) provided an early analysis of the PEP data. Using annual data at the Navy Recruiting District level, he estimated the elasticity of enlistments of highly qualified recruits with respect to all Navy advertising to be about 0.05. That is, doubling Navy advertising would raise

[7]Estimation procedure may account for the different recruiter elasticities estimated by Fernandez (1982) and Dertouzos (1985). Fernandez (1982) used a fixed-effects estimator in his dataset of 67 Military Entrance Processing Stations (MEPS); Dertouzos (1985), who used a 33-MEPS subset of Fernandez's (1982) dataset, did not.

[8]PEP estimates military advertising expenditures and impressions at the county level for many categories of advertising.

TABLE 5-1 Empirical Strategies

Study	Beginning Date of Study	Ending Date of Study	Frequency of Data	Cross-Sectional Unit of Observation	# X-Sec Units
Berner and Daula (1993)	Oct-80	Jan-90	Monthly	Battalion	55
Bohn and Schmitz (1996)	Oct-92	Sep-95	Monthly	NRD	31
Buddin (1991)	Oct-86	Sep-90	Monthly	Battalion	53
Daula and Smith (1985)	Oct-80	Jun-83	Monthly	Battalion	54
Dertouzos (1985)	Dec-79	Sep-81	Monthly	AFEES	33
Dertouzos (1989)	Oct-83	Sep-84	Monthly	ADI	210
Fernandez (1982)	Dec-79	Sep-81	Monthly	AFEES	66
Goldberg (1979)	Jul-71	Dec-77	Quarterly	Nation	1

Services Included in Study	Study Type	Theoretical Framework	Estimation Procedure	Fixed Effects?	Log or Linear
Army	Econometric	Recruiter utility maximization	3-regime switching regression model	Yes	Log
Navy	Navy College Fund	Reduced form	OLS	No	Linear
Army	Army's 2+2+4 Experiment	Recruiter utility maximization	Nonlinear 3SLS	No	Log
Army	Econometric	Supply and demand	2-regime switching regression model	Some models	Log
Army	Econometric	Recruiter utility maximization	2SLS and maximum likelihood	No	Log
All	Advertising Mix Test	Reduced form	SUR with correction for serial correlation	No	Log
Army, Air Force, Navy	Educational Assistance Test Program	Reduced form	12-month first difference using LS with correction for heteroskedasticity	Yes	Log
Navy	Econometric	Reduced form	Maximum likelihood corrected for heteroskedasticity	No	Linear

Continued

TABLE 5-1 Continued

Study	Beginning Date of Study	Ending Date of Study	Frequency of Data	Cross-Sectional Unit of Observation	# X-Sec Units
Hogan et al. (1996)	Jan-90	Dec-94	Monthly	NRD	31
Kearl et al. (1990)	Oct-80	Dec-89	Quarterly	Brigade	5
Murray and McDonald (1999)	Oct-82	Sep-93	Monthly	PUMA	911
Polich et al. (1986)	Jul-81	Jun-84	Monthly	MEPS	66
Smith et al. (1990)	Oct-80	Sep-89	Monthly	Battalion	55
Warner (1990)	Oct-80	Sep-87	Quarterly	NRD	41
Warner (1991)	Oct-80	Sep-90	Annual	NRD	41
Warner, Simon and Payne (2001)	Oct-89	Oct-97	Monthly	State	51

NOTE: ADI = areas of dominant influence; AFEES = armed forces entrance examination station; MEPS = military entrance processing station; NRD = Navy recruiting district; PUMA = public-use microdata areas. FIML = full information maximum likelihood; IV = instru-

Services Included in Study	Study Type	Theoretical Framework	Estimation Procedure	Fixed Effects?	Log or Linear
Navy	Econometric	Reduced form	LS with correction for serial correlation; IV for advertising in some models	Yes	Both
Army	General	Reduced form	GLS heteroskedasticity	No	Log
All	Econometric	Hybrid structural and reduced form	OLS corrected for heteroskedasticity and serial correlation; IV for some variables	Yes	Log
Army	Enlistment Bonus	Recruiter utility maximization	Two-stage procedure using 3SLS	No	Log
Army	Econometric	Enlistee utility maximization	OLS—found that correcting for serial correlation did not affect estimates	Yes	Log
All	Econometric	Reduced form	Effects	Yes	Log
Navy	Econometric	Recruiter utility maximization	OLS and fixed effects	Yes	Log
All	Econometric	Recruiter utility maximization	Fixed effects with IV for some variables	Yes	Both

mental variables; LS = least squares; GLS = generalized least squares; OLS = ordinary least squares; 2SLS = 2-stage least squares; 3SLS = 2SLS followed by SUR; SUR = seemingly unrelated regressions.

TABLE 5-2 Econometric Estimates of Advertising Elasticities

Study Sample
Berner and Daula (1993)
Bohn and Schmitz (1996)
OLS
NRD dummies included
NRD and month dummies included
Buddin (1991)
Daula and Smith (1985)
"Pooled"
Supply-Constrained
Demand-Constrained
Dertouzos (1985)
Reduced form, 1980—goals included
Reduced form, 1981—goals included
Structural model 1980—2SLS
Structural model 1981—2SLS
Structural model 1980—FIML
Structural model 1981—FIML
Dertouzos (1989)
Army
Navy
Air Force
Marines
Fernandez (1982)
Army
Navy
Air Force
Goldberg (1979)
Hogan et al. (1996): Median estimates
TV
Radio
Mailings

Elasticities			
Service Advertising	Joint Advertising	Recruiters	Advertising Measure National Impressions
0.208	NA	0.274	
NA	NA	0.221	
NA	NA	0.346	
NA	NA	0.139	
NA	NA	0.238	Expenditures
			Impressions and expenditures
0.089	NA	0.585	
0.107	NA	0.959	
0.156	NA	0.826	
NA	NA	0.842	
NA	NA	0.466	
NA	NA	1.193	
NA	NA	1.086	
NA	NA	1.647	
NA	NA	1.529	
			Expenditures
0.028	0.016	0.227	
–0.005	0.028	0.526	
0.071	0.008	0.303	
–0.001	0.023	0.470	
		0.295	
		0.274	
		0.090	
0.140		1.270	Dollars
		0.286	
0.028	0.031		Impressions
0.021	0.009		Impressions
0.038	0.029		Impressions

Continued

TABLE 5-2 Continued

Study Sample
Kearl et al. (1990)
Model 1
Model 2
Model 3
Murray and McDonald (1999)
Army early (1983-87)
Army late (1990-93)
Marine Corps early (1983-87)
Marine Corps late (1990-93)
Air Force early (1983-87)
Air Force late (1990-93)
Navy early (1983-87)
Navy late (1990-93)
Polich et al. (1986)
Smith et al. (1990)
Warner (1990)
Army, time trend included
Navy, time trend included
Air Force, time trend included
Marine Corps, time trend included
Warner (1991)
Warner, Simon, and Payne (2001)
Army
Navy
Air Force
Marine Corps
SUMMARY STATISTICS
Mean
Standard deviation
Coefficient of variation

NOTE: 2SLS = 2-stage least squares; FIML = full information maximum likelihood; NRD = Navy recruiting district; OLS = ordinary least squares.

Elasticities			
Service Advertising	Joint Advertising	Recruiters	Advertising Measure National Impressions
			Expenditures
0.430		0.480	
0.580		0.680	
0.720		1.150	
		0.51	
		0.60	
		0.53	
		0.62	
		0.49	
		0.59	
		0.33	
		0.24	
0.056		0.597	Expenditures
0.050		0.150	Expenditures
			Expenditures
0.103	0	0.371	
0.015	–0.004	0.412	
–0.034	0.004	–0.045	
–0.017	0.001	0.487	
0.050	–0.028	0.527	Expenditures
0.136	0.008	0.410	Impressions
0.084	–0.003	0.640	Impressions
–0.013	0.015	0.480	Impressions
–0.065	0.022	0.470	Impressions
0.114	0.010	0.551	
0.186	0.015	0.368	
1.625	1.545	0.667	

enlistments of highly qualified recruits by 5 percent. Despite the apparently low response, advertising was found to be cost-effective in comparison to recruiters because Navy advertising was a very low part of the overall recruiting budget in the 1980s (see the cost estimates section below).

Hogan et al. (1996) examined the impact of Navy advertising with more recent data. Furthermore, they estimated the effects of various subcategories of advertising. Using a fixed-effects model, they estimated a TV elasticity of 0.03, a radio advertising estimate of 0.02, and a magazine advertising elasticity of 0.04. Elasticities were also estimated for joint-Service TV (0.031) and radio advertising, and mail advertising (own Service, 0.038; joint-Service mail, 0.029).

Warner, Simon, and Payne (2001) utilized PEP data to estimate an overall advertising elasticity for the Army of 0.13 and for the Navy of 0.08. In models that separated advertising into TV and non-TV advertising, they obtained TV elasticity estimates of 0.09 and 0.05 for the Army and Navy, respectively, and non-TV estimates of 0.07 and 0.05 for those Services. But no relationship was found between Air Force and Marine Corps enlistments of highly qualified recruits and advertising. Those Services' advertising programs are much smaller than the Army and Navy programs, so the insignificance of the estimates could reflect the smaller scale of those programs. Furthermore, there was some doubt about the quality of the data for those Services.

Because these studies all imposed log-log or semilog functional relationships between *log(H)* and advertising, one should not use these estimates to infer the effects of advertising over a wide possible range of advertising expenditures. Dertouzos and Garber (2003) recently reanalyzed the Ad Mix Test data imposing the logistic relationship between *log(H)* and four forms of advertising: TV, radio, magazine, and newspaper. Estimates were consistent with the theory and suggested different S-shaped curves for the different media types. Newspaper spending did not appear to have an impact on *log(H)* at any level of advertising. Magazine advertising had an effect at a very low level but reached the saturation level at a very low level of spending. Radio advertising had larger minimum effectiveness and larger saturation levels than magazine advertising. TV advertising had the largest minimum effectiveness and the largest saturation levels.

Dertouzos and Garber report extreme difficulty in estimating their models with more recent (fiscal year 1993-1997) data.[9] Service-specific estimates were not, in their opinion, very reliable. They therefore aggre-

[9]In particular, the logistic advertising model must be estimated with nonlinear regression, and they had difficulty getting nonlinear regression procedures to converge. And when they did, estimates did not seem to be very plausible.

gated their data to the DoD level and estimated models of total DoD enlistments as a function of advertising and other control variables. Some models contained total advertising, while others contained TV and non-TV advertising. These models yielded significant positive effects of total advertising and of TV advertising, but not of non-TV advertising.

Recruiting Cost Function

The concept of recruiting cost functions (RCF) is taken from the economic theory of production. The RCF provides an estimate of the optimal (i.e., cost minimizing) resource levels to achieve a given set of recruiting goals. They can be derived directly from the enlistment supply models, which may be interpreted as recruiting production functions.

The RCF is derived as the outcome of the following cost-minimization problem:

$$\text{choose } X \text{ to } \min C = \textstyle\sum_{i=1}^{n} p_i X_i \text{ subject to } f(H^*, L^*, X, Z^*, E) = 0.$$

where the p's are the prices or unit costs of n resources over which there is choice, the X's, and H^* and L^* are particular values for highly qualified and less-well-qualified recruits, respectively. The vector Z consists of factors over which the Services do not have control but that affect recruiting. It includes the civilian unemployment rate and civilian wages as well as, arguably, the aggregate military pay raise. The vector X includes recruiters, bonuses, advertising, and college benefits, among other things.

As a result of solving the minimization problem and solving the first order equations for a minimum, we obtain the recruiting cost function:

$$C = C(p,H,L,Z)$$

where C is the minimum total cost of recruiting H highly qualified recruits and L recruits with low qualification levels, and p is a vector of resource prices and Z are factors affecting cost that are beyond the control or choice of the Service. In addition to providing an estimate of the minimum total cost, the RCF also provides an estimate of the optimal levels of resources that constitute the cost. That is, a product of the cost function is an estimate of the optimal amount of each resource, given the overall goals, where "optimal" is the cost-minimizing amounts. Moreover, by differentiating with respect to H or to L, one obtains an estimate of the marginal cost of highly qualified and less-well-qualified recruits.

A RCF was developed for each Service by Hogan and Smith (1994). A version of the RCF is currently in use by the Office of the Secretary of

Defense and the Chief of the Naval Recruiting Command. In these applications, the RCF uses a log-log specification of the underlying enlistment supply model. The model provides the cost and levels of resources for a given set of recruiting goals and for a given economic environment.

Marginal Resource Cost Estimates[10]

Once the responsiveness of enlistment to recruiting resources has been estimated, estimates can be used to calculate the marginal cost of enlistment. A greater responsiveness of enlistment to recruiting resources implies lower marginal cost. Consider the marginal cost of highly qualified enlistments brought about by an expansion of the recruiter force. The marginal cost of highly qualified enlistment via recruiters can be calculated as $C^*(R/H)/\eta_R$ where C is the cost of a recruiter, R is the ratio of recruiters to highly qualified enlistments, and η_R is the elasticity of H with respect to R. If $C = \$45{,}000$ (DoD recruiter cost factor), $R/H = 0.1$ (late 1990s ratio for the Army, the Navy, and the Marine Corps), and $\eta_R = 0.5$, then the marginal cost of H via additional recruiters would be $9,000. Since η_R is 0.5, the marginal cost is twice the average cost ($4,500). The most plausible estimates of η_R range from 0.3 to 0.6 (Table 5-2). This range of estimates implies marginal recruiter costs for the Army, the Navy, and the Marine Corps in the range of $7,500 to $15,000. Because Air Force recruiters average about 25 highly qualified contracts per recruiter, marginal recruiter cost for the Air Force is much lower (about $3,400).

The marginal cost of highly qualified enlistments brought about by an expansion in advertising can be calculated similarly. At 1997 budget levels, Warner, Simon, and Payne calculated the marginal cost of recruits via an expansion of Army advertising to be about $10,700. Because they

[10]Marginal cost estimates reported here are inclusive of "rents." Rents are the amounts over and above that amount necessary to obtain the recruits. They are more important for some ways of obtaining additional recruits than others. For example, additional recruits will be attracted to enlist if there is an increase in first-term pay. But because all recruits, even those who would have enlisted without the increase in first-term pay, would receive the pay increase, rents from this method of increasing recruits supplied are large. In contrast, rents associated with increasing recruits through the efforts of additional recruiters are negligible. From the perspective of the economy as a whole, rents are approximately a transfer payment, neither a cost nor a benefit. They represent resources transferred from some to others in the economy, although there may be additional costs if the resources are taxed from some, and this taxation affects behavior. Nonrent costs are always costs to the economy. They represent resources used up in the production of some good or service. Hence, if two or more ways of increasing the supply of qualified recruits have the same marginal cost when rents are included, the one with the lower marginal cost excluding rents is preferred from the perspective of the economy as a whole.

estimated a larger responsiveness of enlistment to advertising than those obtained in some of the studies summarized in Table 5-2, marginal advertising costs could in fact be greater.

The logistics functional form employed by Dertouzos and Garber implies high marginal advertising costs at low levels of enlistment followed by declining marginal costs in the midrange of advertising outlays and then by high marginal costs as advertising reaches a saturation point (which may depend on the target market and the advertising message strategy). Dertouzos and Garber calculated that, at 1993-1997 advertising budget levels, the marginal cost of a DoD highly qualified male enlistment via total advertising to be about $37,000. But FY 1993-1997 was a period of relatively low advertising; advertising outlays rose dramatically in the FY 1998-2000 period. Dertouzos and Garber (2003) calculated that, at an advertising spending level roughly double the average level prevailing in the FY 1993-1997 period, marginal advertising costs for highly qualified male contracts would fall to about $10,500. Furthermore, according to their calculations, marginal costs of the recruits obtained via TV advertising would continue to fall over a much higher range of spending before beginning to increase.[11]

Estimates of the Effects of Other Resources

Econometric methods can be applied to estimate the effects of other resources not often considered in the traditional enlistment supply model. An example is an econometric estimate of the effect of the number and location of recruiting stations on recruit supply (Hogan, Mehay, and Hughes, 1998). In this analysis, an enlistment supply model was estimated that was not dissimilar from the models described above. It consisted of a pooled time series of cross-sections. The dependent variable was specified as enlistments in a given zip code area. Explanatory variables included, among other things, distance to the nearest recruiting station. This permitted assessment of the effect of recruiting station location on enlistment supply. Estimates indicated that the number and location of recruiting stations had a significant effect on recruiting.

Econometric methods can often be applied to include additional resources or factors that could potentially affect recruiting. The data for the additional resource or factor must be available for the same cross-

[11]These cost estimates are dependent on the assumed functional (logistic) form. While certainly the most plausible functional form found in the recruiting literature, more work clearly needs to be done to verify that it fits the data better than other forms. Given the difficulty of estimating the model and the probable fragility of their estimates, Dertouzos and Garber stress the need for more work in this area.

sectional and time-series dimensions as the other variables included in the model. What is important is that the effect of the additional variable be included in a model that also includes all other resources or factors affecting recruiting. Failure to include all other variables increases the likelihood of obtaining a biased estimate of the effect.

IMPROVEMENTS IN DOD RECRUITING RESEARCH

Effects of Recruiters

Aspects of the econometric estimates of the effects of recruiters on enlistment supply can be improved by a relatively straightforward extension of existing models. First, an individual recruiter's productivity varies with experience. Newly assigned recruiters typically produce few or even no recruits in their first six months on station. Over the next 18 months, productivity rises and reaches a plateau. Then productivity begins to decline as the recruiter prepares to rotate back to his or her primary skill area.[12] Hence, one can expect the productivity of recruiters to vary systematically with experience. When the recruiting force increases significantly in a short period of time, average experience declines. Similarly, when the recruiting force declines rapidly, it is usually by a disproportionate decline in new recruits, increasing average experience and productivity.

Failure to account for the effect of these changes on the average experience and average productivity of recruiters as a whole may bias the measured recruiter productivity toward zero. By measuring the average experience or tenure of recruiters in the estimation equation, it may be possible to improve the precision of estimates and perhaps eliminate a bias in the estimation of recruiter effects on supply. Moreover, by estimating econometric models for which the dependent variable is a measure of *individual* recruiter productivity, insights can be gleaned regarding factors affecting individual productivity.

A second issue in measuring the effects of recruiters is to expand the analysis of recruiter incentives beyond the effects of aggregate quotas. This would attempt to capture econometrically the more sophisticated Service "point" systems and other complex incentive structures (see Asch, 1990, for an early analysis of the Navy recruiter point system).

Finally, a third issue that has not been addressed in the literature is to include the effects of reserve force competition on active-duty recruiting. While other Service competition has been included in several econometric

---

[12] This is described for a cohort of recruiters in McCloy et al. (2001)

specifications, no models have incorporated reserve force component competition for nonprior service recruits. For the selected reserve force, the proportion of nonprior-service recruits is likely to rise relative to those who affiliate while leaving active-duty service, simply because active strength has declined and retention has increased. This may foster greater competition with active recruiting, in that in many instances they are competing for recruits from the same set of high school seniors and others. Incorporating reserve force recruiting into active-duty recruiting models may help explain the potential interactions, while providing estimates of the factors affecting reserve force recruiting.

Advertising Content

Econometric estimates that incorporate the effect of advertising have measured advertising largely as homogenous counts of "impressions"—the number in relevant populations who see or hear the advertisement—or expenditures by period and geographic location. They have not attempted to get inside the expenditure or impression to measure differences in effects by specific advertising content.

Data could be constructed for television advertising that distinguished the content of the advertisement as well. That is, if there are one or two dominant advertising themes for a campaign for a given Service in time period 1, but this changed in time period 2 and again in time period 3, the advertising variable itself could be constructed to allow for different effects for the same ostensible level of advertising (impressions or expenditures) over each of these three periods.

Previous econometric models have separately distinguished the effects of various types of advertising on recruiting. Radio, magazine, or print advertising and television advertising have been distinguished, for example. If the time periods for major television advertising themes or campaigns can also be distinguished in the data, it is not difficult to specify the model to allow them to have separate effects. Distinguishing different effects for television advertising based on the content of the advertisements themselves, however, would be difficult because one would be attributing a portion of the variation over time in recruiting to the shift among advertising themes or specific advertisements. The model specification and supporting data must be solid to isolate the effects.

An alternative approach to allowing separate estimates by advertising campaign would be to try and isolate essential elements of an advertising theme and measure them in continuous variables. This would undoubtedly entail some subjectivity. One could, however, classify ads by the seconds devoted to three or four themes—training, postservice education, compensation, adventure, and patriotism, for example. This would be an interesting exercise, but it would not distinguish differences in the quality of a given message within a given theme. That is, it would not distinguish between good advertisements that feature training as a theme from bad advertisements that feature training as a theme.

Hence, we suggest that initial attempts in this area try to test for differences in effects of advertising campaigns by allowing distinguishable campaigns to have different effects. Then, if it is established that effects do appear to vary significantly by content and not just dollars or expenditures, a second step would be to try to understand how content variation affects recruiting.

More Flexible Functional Forms

A limitation of much of the empirical research is that the functional forms of the econometric specifications have been relatively restricted. Perhaps one way to understand some of the implications of the particular functional forms is to consider an enlistment supply equation as analogous to a production function for recruits. The log-log form of the supply curve that has been popular forces elasticity estimates to be constant, regardless of level of recruiting activity or of the particular recruiting factor. Moreover, when the RCF is derived from such estimates, it is readily seen that the factor shares of each price resource—the proportion of total cost of production attributable to each resource—are constant regardless of how the prices of resources change.

More flexible functional forms would provide less restrictive constraints on parameter estimates. In the literature reviewed, this is particularly important for the measurement of the effects of advertising and for modeling the effects on recruiter incentives. But it is also important to obtain more precise and useful estimates of the effects of resources in general. The more flexible functional forms, such as the trans-log formulation, typically require richer data for estimation. This is in part because these forms allow the data to determine whether the effects of factors may vary with scale or with relative proportions of other factors, rather than imposing them as constraints in the mathematical formulation. As panel data become more refined and become available over longer periods, application of more flexible functional forms can perhaps provide new sets of insights on factors affecting recruiting.

A theme underlying all of the suggested areas for improvement is the need for better data, consistently collected and retained over time. Suggested improvements in estimates of the effects of advertising, recruiters, and other incentives will require more sophisticated functional forms and more detailed specification of effects. This places greater demands on data than has historically been the case.

Ideally, these data should include enlistment contract and accession data, by level of qualification and Service, at the lowest reasonable level of aggregation and time period. Data at the individual recruit level, coupled with information regarding timing of enlistment, home (zip code) of the recruit, and recruiting station should be maintained over time. Importantly, the data should also include information on the resources and incentives that have been applied and the external factors that were in effect during the period, as well as indications of recruiting policies, incentives, quotas, and so forth. Advertising data, which we discuss in several places, should contain information not simply on impressions or dollar expenditures, but also include a systematic characterization of advertising content, if these are to be evaluated. These data should, again, be able to be tied to the recruiting data at the lowest reasonable level of aggregation.

6

The Timing and Levels of Joint and Service-Specific Advertising

In this chapter we discuss research regarding two questions. First, is there a minimum level of advertising necessary for a cost-effective recruiting program, even if that advertising is not necessary to achieve contemporaneous enlistment contract goals? Historically, when the recruiting climate is good and recruits are plentiful, military planners tend to cut advertising budgets, thereby contributing to a reduction in "awareness" capital and propensity levels. This may possibly set up a boom or bust cycle, in which propensity falls, recruiting becomes more difficult, and then advertising funds have to be restored—or even increased beyond what they would have been without the initial cut—to stimulate propensity and enlistment. To our knowledge, this issue has not been studied extensively, and therefore the best research designs may not be immediately clear.[1] This chapter discusses the possibility of both econometric and experimental designs to address the question.

The second question concerns the proper levels of joint and Service-specific advertising. Certain types of advertising themes, such as generic themes designed to increase overall propensity, may be best done as a joint program, while advertising themes featuring specific benefits of mili-

[1]Goldberg and Kimko (2003) studied aggregate enlistments into the four Services over time and estimate larger business cycle (e.g., unemployment) effects for the Army and the Navy than are generally estimated in the studies summarized in Chapter 5. They recommend an early warning system that will help the Department of Defense (DoD) predict turning points in recruiting and undertake policy actions that will reduce the cyclicality of recruiting.

tary service are best done in the Service programs. We do not currently know much about what levels of Service and joint advertising will most efficiently sustain different levels of propensity and enlistment. Answering this question is most likely to require a combination of research designs, both econometric and experimental.

MINIMUM LEVEL OF ADVERTISING

In the discussion in Chapter 5 of the optimal levels of recruiting resources to achieve a given set of recruiting goals, the optimal amount of advertising, like the optimal amount of any other recruiting resource, was the amount that permits goal achievement at lowest cost. Is there some minimum level of advertising that is above the level of advertising that minimizes the total cost of achieving current recruiting goals, but may be optimal in a longer term perspective? For example, consider a case in which, because of a poor economy and high unemployment rates, or a significant reduction in the current demand for new recruits, or both, current recruiting goals can be achieved even if recruiting resources, including advertising, are drastically reduced. Is there a reason to maintain resources, particularly advertising, above the minimum level to achieve current goals?

If the answer is yes, it must be because greater advertising expenditures in one period will make recruiting goals in a future period less costly to achieve. That is, there must be dynamic aspects to some types of recruiting resources such that current expenditures on those resources affect both current and future recruiting. As we discussed in Chapter 5 and as reported in the literature (Hogan, Dali, Mackin, and Mackie, 1996; Dertouzos and Garber, 2003), current advertising expenditures contribute to a stock of information and awareness capital. This stock changes over time, as the existing stock of information decays but is replenished by new advertising expenditures.

Hence, under this basic concept of a flow of advertising contributing to a stock of awareness or information capital, one can anticipate that the effect of reducing the flow of new advertising on recruiting will, at first, be small. But as the stock continues to decay without replenishment, the adverse effect on recruiting can grow. This is illustrated in Figure 6-1, which assumes a depreciation rate of 10 percent. The stock of information and awareness capital has been built up to a notional value of 100 units in month 1. However, from that month forward, new advertising, which contributes to the stock, is cut to zero. Some level of new advertising is necessary simply to compensate for the depreciation of the existing stock. The effect on the ability to recruit is, at first, modest, because the stock of capital by month 3 is about 80 percent of the original stock. However,

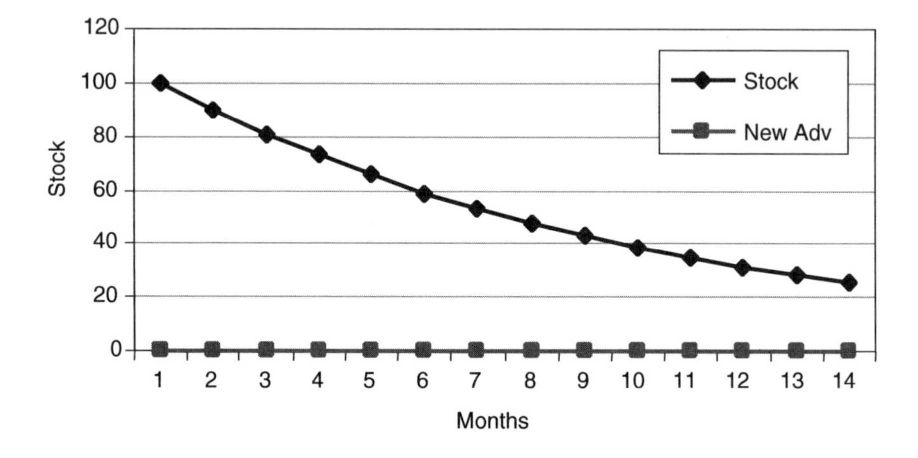

FIGURE 6-1 Stock of information and awareness capital at 10 percent monthly depreciation (notional).

with no new advertising and continued depreciation, the stock would quickly be depleted to less than 30 percent of its original value.

Now, assume that new advertising, equivalent to a contribution to the stock of 15 units per month, is provided. In this case, the stock of information, initially at 100, rises eventually to a new equilibrium of 150, where depreciation (10 percent of 150) equals new advertising of 15 units. At this point, new advertising exactly offsets the depreciation, maintaining a stock of information and awareness capital of about 150. When new advertising exceeds depreciation, the stock grows. This is shown in Figure 6-2.

This simple example suggests how one may reduce advertising resources and in the short term experience no significant reduction in recruits. Moreover, assume that, initially, the optimal level (i.e., cost-minimizing level) of the stock of advertising is 100. Then, because the economy sours and the demand for recruits declines, the new optimal level of the stock is 50. That is, significantly less awareness is necessary to recruit the desired numbers. In this case, it would make sense to reduce new advertising until the new equilibrium stock is approached, then increase it again.

However, if there is reason to believe that external economic conditions are likely to improve significantly, or that the demand for new recruits is to increase, it may be imprudent to permit the stock of awareness capital decline to this level. This would be of greater concern if the marginal cost of increasing awareness capital in a particular period was

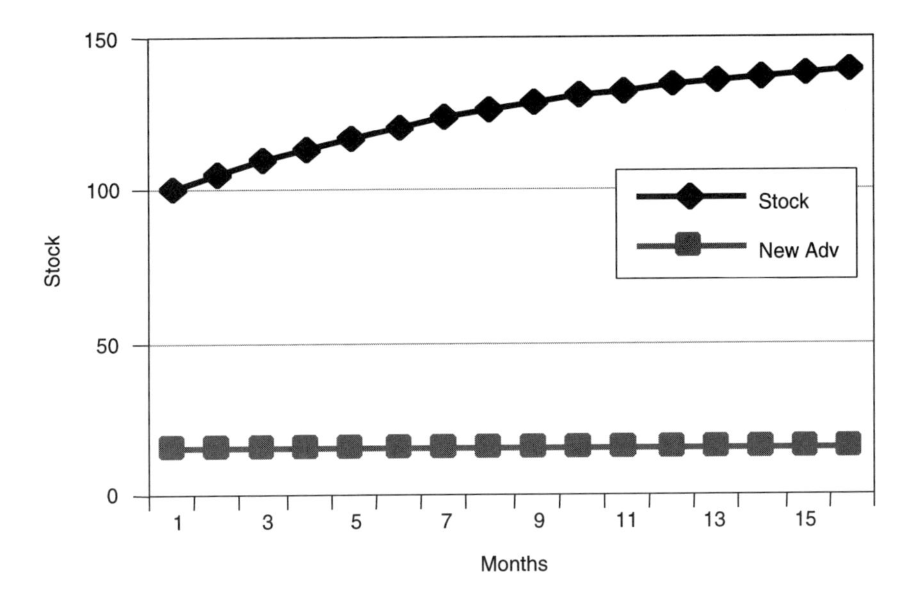

FIGURE 6-2 Stock of information and awareness capital (notional).

increasing. That is, let A be the augmentation to the stock of awareness capital in a given time period. Then, let $C(A)$ be the cost of increasing awareness capital by A in a given period (month). We have:

$$\partial C(A) / \partial A > 0$$
$$\partial^2 C(A) / \partial A^2 > 0$$

If the marginal cost of increasing the stock of awareness capital increases with the magnitude of the desired increase in capital, then it may be less costly to begin increasing the stock of information capital in periods earlier than is needed for current recruiting.

To abstract from uncertainty over the demand for recruits, let us assume that the time path of recruit quotas for both highly qualified recruits (Q_H) and less-well-qualified recruits (Q_L) is known and that it fluctuates over time. One way to express the issue is whether there is a case for which

$$\sum_{t=1}^{T} C_{\min}(Q_{H,t}, Q_{L,t}, p_t, E_t) / (1+r)^t > \sum_{t=1}^{T} C_{opt}(Q_{H,t}, Q_{L,t}, p_t, E_t) / (1+r)^t$$

where *C(....)* is the cost function for producing highly qualified and less qualified recruits, respectively, in period t, and r is the discount rate. The variable P is the vector of recruiting resource prices in the period and E is a vector representing the state of the economy. The notation *min* represents the cost-minimizing levels of inputs to exactly produce the supply of recruits to meet goals in period t, while *opt* is the level of resources that minimizes, not the costs in a specific period, but the present value of costs over periods 1 through T, when discounted at interest rate r. In the second case, potential supply may be greater than necessary to meet current goals. If so, it is optimal because it lowers the costs of meeting subsequent goals. If the above inequality holds as a general case, then there is a level of resources above the level necessary to meet current goals that minimizes the present value of costs over the longer term.

How can we test the hypothesis that the inequality, defined above, may hold? We are likely to be able to do so only indirectly. If advertising is the reason the inequality holds, it is necessary that: (1) advertising expenditures have effects over time and (2) the marginal cost of increasing the stock of information or awareness capital in a given time period is increasing.

If advertising affects the stock of information capital only in the period in which it is produced, then the stock of capital and the flow of advertising in a given period are equal, except for any external factors affecting the stock. In a sense, the depreciation rate of the inherited stock is 100 percent. The only reason to increase advertising in a given period would be to affect current enlistment supply, and there would be no reason not to reduce advertising if not needed in the current period. Furthermore, if advertising adds to the effective stock of information at a constant marginal cost, it would be less costly, in a present value sense, to wait until the advertising is needed before increasing it, and to do it all in one period. Only if marginal costs are rising, and advertising effects are dynamic, would it be optimal (i.e., cost minimizing) to incur advertising costs currently, not because they are needed currently but because they affect future recruits.

To determine if there is a minimal level of advertising expenditure that is above that necessary to meet current goals, an econometric model must be able to capture the effects of advertising over time and allow for nonlinear effects of advertising on recruiting in a given time period.[2]

---

[2]Earlier we presented the notion that advertising contributed to a stock of information or awareness capital, and that this stock then affected recruiting. This is a useful concept for exposition purposes. However, the nonlinear effect, in practice, is between advertising in a given period and enlistments in that and subsequent periods. There is no necessity for the concept of the stock.

Dertouzos and Garber (2003) estimate a supply curve that includes advertising. As discussed in Chapter 5, this specification permitted both dynamic effects of advertising—that is, it allowed advertising in one period to affect recruiting not only in the same period but in subsequent periods—and nonlinear effects within a given time period. Advertising in each period, both current and lagged, enters as a nonlinear S-shaped or logistic function. The functional form provides ranges for small marginal effects of advertising, then increasingly larger marginal effects, followed again by increasingly small marginal effects, as if there were a saturation level for advertising. If the unit price of advertising is constant, this would correspond to areas of high but decreasing marginal costs, followed by an inflection point and increasing marginal costs. Hence, this specification has the ability to capture the two features of advertising that would be required if there is a minimum level of advertising that is above the cost-minimizing level when only current recruiting goals are considered.

The findings reported in Dertouzos and Garber (2003) and those from the literature reviewed briefly in Chapter 4 are that although advertising has dynamic effects, the lag structure suggests that most effects are realized within about two to three months of the advertising expenditure. These data suggest that current advertising may affect recruiting one, two, or three months in the future, but it does not affect recruiting six months or a year in the future. If the data on which these findings are based are strong and compelling, it would mean that a compelling case could not be made for maintaining advertising expenditures in a given period, if they are not needed to achieve current goals, solely because it is desirable to increase recruit supply six months or a year in the future. However, there is not a large body of evidence on the issue, and most authors who do provide some evidence also note that the data from which the estimates are derived are weak, and that more research is necessary. Hence, because of the paucity and limitations of evidence available to date, we conclude that the issue remains open and subject to further research. In our view it would be a serious mistake to view the available research as sufficient grounds for drawing conclusions one way or another about the effects of current advertising on future outcomes.

Furthermore, the functional form used by Dertouzos and Garber (2003), while allowing for nonlinear effects of advertising—a necessary feature for establishing some minimal level of advertising—constrains the nonlinearity to follow a particular form. Because of this, it imposes the nature of the nonlinearity, rather than having a very flexible functional form that would allow the data to determine the nature of the nonlinearity. Hence, while the S-shaped functional form for advertising is plausible, it cannot be used by itself to test the hypothesis that advertising has nonlinear effects on recruiting.[3] Instead, a more flexible functional form, one that permits lagged effects of advertising but also nonlinear effects of advertising within a period and does not constrain the precise nonlinear relationship, should be applied.

The data necessary to estimate more flexible functional forms are typically quite demanding. Because the functional form itself places fewer constraints on the nature of the relationship between the supply of recruits, recruiting resources, and other factors affecting recruiting, the variation in the data must be sufficient to support distinguishing among alternative relationships. It is probably not the case that the most recent data over the 1990s and early 2000s are sufficient to estimate these more demanding relationships. Dertouzos and Garber (2003), for example, were not able to estimate a Service-specific enlistment supply function from data over the 1990s, even though they placed constraints on the nature of the functional relationships.

A controlled experiment using a quasi-experimental design to generate the data necessary to estimate supply relationships that incorporate the potential for both dynamic effects of advertising and nonlinear relationships is likely to be impractical. The reason is that the effects to be estimated are not simple impact effects of an intervention, but a more complicated set of relationships regarding the dynamic, nonlinear structure of advertising effects. To generate sufficient data to estimate such relationships would require that the experiment be continued over an extended period of time—at least three or four years. In an operational environment in which there are changing supply conditions and real recruiting demands to be met, this is unlikely.

Instead, we propose a focused effort to maintain data, especially advertising data, in a systematic and careful way, for the purposes of estimating a supply curve that incorporates the potential both for dynamic and nonlinear advertising effects. This is consistent with our recommendation in Chapter 5 for better and more thorough data collection and estimation using more flexible functional forms.

Finally, it is important to note that most recruiting advertising has been targeted directly on recruit-age youth. Advertising designed to influence parents, others who counsel youth, and youth who are not yet old enough to enlist may have effects that extend over time. Tests for the dynamic effects of advertising should attempt to distinguish between

---

[3]It could be used in conjunction with other constrained forms to test the hypothesis. In that case, one would specify several alternative forms, including a linear form, and determine which form explained the observed data the best, and which could provide the best out-of-sample prediction.

advertising targeted at recruit-age cohorts and advertising targeted at influencers or those who are not yet of recruiting age, if possible.

ADVERTISING TARGETED AT THE LONGER TERM

Historically, advertising for recruiting has focused almost exclusively on recruit-age cohorts of youth. That is, it has been focused on providing 17-21-year-olds with information about opportunities in the Services and on convincing recruit-age youth that enlisting is an option that they should seriously consider. This being the case, it is perhaps not surprising that the literature finds that current advertising expenditures affect recruiting only in the near term—two or three months in the future—and do not have effects on the market in the longer term—six months or a year or more in the future. An alternative type of advertising would target audiences that have, perhaps, a longer time horizon for affecting the recruiting market. This would include targeting youth earlier in the career decision process—before they can actually enlist, say at ages 14 through 16—and targeting parents and other adult authority figures who may influence the decisions of youth both immediately and in the future.[4]

This type of advertising has not been systematically employed by the Services or the Department of Defense. Arguably, it is a possible role for joint advertising. There is a solid logical argument for this type of advertising, based on analyses of survey and other data (see National Research Council, 2002, for further discussion). However, because it has not been tested in the recruiting market, little is known about its effects.

The first step toward understanding its potential effectiveness and how it may influence the recruiting market over time would be to develop advertising of this nature and test it in the recruiting market. Because we anticipate that this type of advertising has the potential for affecting the market over longer periods of time, the experiment should be designed to devote advertising expenditures of this nature in the recruiting market for an extended period of time—at least two years.

Ideally, the quasi-experimental design should attempt to include variation over time and cross-sectionally in the advertising test. However, in an operational environment, a simple pre-post design may be both

---

[4]Warner, Simon, and Payne (2002) provide evidence that military advertising affects propensity to enlist. In an analysis of the Youth Attitude Tracking Study propensity over the 1988-2000 time period, they found a youth's propensity to enlist to be positively related to total military advertising expenditures per youth in the youth's state of residence in the 12 months prior to the survey. But data limitations prevented determining whether the impact of advertising on propensity varies by media or by whether it is joint or Service-specific advertising.

more workable and may serve to deliver the advertising in a more efficient way.[5] If the experimental period is of sufficient duration, it would be useful to vary the level of advertising over time, rather than keeping it at a single level "dose." In addition to econometric analyses of effects, one would also include more qualitative methods of evaluation, including surveys and focus groups.

THE LEVELS OF SERVICE-SPECIFIC AND JOINT ADVERTISING

In fiscal year (FY) 1989, the last large recruiting year before the drawdown, the Services spent $100 million on enlisted advertising compared with a joint level of $20 million. In that year, almost the entire joint budget was for TV advertising. By FY 2000, joint advertising had dwindled to less than $5 million despite Service advertising of $250 million (Office of the Assistant Secretary of Defense [Force Management Policy], 2000). The scaled-back joint program has been either direct mail or magazine advertising. The joint program raises four fundamental questions:

- What should be the content of joint advertising, and should it differ from the content of Service-specific advertising?
- To whom should it be targeted?
- What media should the joint program use?
- Do the efficient levels of joint and Service-specific advertising depend on the scale of the recruiting effort?

As discussed below, the answers to these questions are interrelated. Past research is of little use in answering them. The first two questions have barely been addressed in previous research (Carroll, 1987). Several econometric studies have tried to answer the latter two questions. Here the research strategy has been to estimate the responsiveness of enlistments to Service-specific and joint advertising in different media and then infer the cost-effectiveness of the various forms of advertising in generating enlistments. Evidence from such studies about the relative effectiveness of joint advertising is, at best, mixed (see Table 5-2). Hogan et al. (1996) found that joint TV and joint direct mailing were as effective as Navy advertising in the same media. But estimates by Dertouzos (1989), Warner (1991), and Warner, Simon, and Payne (2001) are less kind to the joint program.

---

[5]Excluding geographic areas to induce cross-sectional variation in advertising is difficult and likely to be cumbersome to recruiting operations. Moreover, it may preclude some efficient ways of delivering advertising, such as national magazines and network television.

Dertouzos and Garber (2003) question the validity of past estimates of the effects of advertising (summarized in Chapter 5). They argue that enlistments follow an S-shaped relationship with advertising and that the form of the relationship varies by media. If their analysis is correct, then we actually know very little about the effectiveness of joint advertising (and therefore the answers to the above questions). In fact, the insignificance of the joint program in past econometric studies may have resulted from its being below the minimum efficient scale (MES) in Figure 5-1.

Despite criticizing past studies on theoretical grounds, estimates provided in Dertouzos and Garber (2003) do not help answer the above questions. Their updated analysis of the Ad Mix Test data does not distinguish joint advertising from Army advertising, their analysis of the 1993-1997 recruiting experience is conducted at the DoD level, and the joint program is added to Service-specific advertising to derive measures of total DoD advertising. Since past work is of little use in answering the four fundamental questions above, this section now explores a series of questions that need to be addressed in order to adequately answer these fundamental questions. It is apparent from this discussion that an experiment is needed.

Advertising Content

If the joint program advertises in the same media as the Service programs with the same message content, it is just an add-on to Service-specific advertising. That is not necessarily bad. Joint advertising that mirrors the Service programs essentially expands the scale of a common program and thereby just moves enlistments up the S-shaped curve portrayed in Figure 5-1. If the unit cost of joint and Service-specific advertising are the same, joint advertising that mimics the message and content of Service advertising would be a perfect substitute for Service advertising. Such advertising would be pointless since an equivalent expansion of the Service programs would yield the same result. In fact, due to the overhead costs of ad development and effects of program scale on advertising rates negotiated with ad agencies, newspapers, and other media, it is unlikely that the joint program could perfectly replicate the Service programs at the same cost. In order to be useful, joint advertising must do something that Service-specific advertising does not.

One way for joint advertising to distinguish itself is in the content of its messages. The committee's previous report (National Research Council, 2003, p. 227-230) discussed the message strategies in current military advertising. Service-specific advertising appears to be occupationally oriented, career-oriented, or oriented toward appealing to youths' desires for adventure, challenge, self-development, and self-actualization. The report labeled these "package-oriented" advertising and went on to note that only Marine advertising seems directed toward the more noble virtues of patriotism, self-sacrifice, and service to country. (This seems to have changed with recent Service advertising in the wake of September 11, 2001, and Operation Iraqi Freedom.)

Two recent DoD study groups—the Defense Science Board Task Force on Human Resources Strategy (DSB) and the DoD Quality of Life Panel (also called the Jeremiah panel)—expressed a belief that DoD needs to do more to engage the American public about the importance of public service.[6] More values-oriented advertising that stresses the virtues of patriotism, self-sacrifice, and service to country would serve that purpose. Of course, the ultimate goal of advertising is to make youth more inclined to serve (i.e., increase propensity), so that more of them walk into recruiting offices on their own and more respond positively when they are contacted by military recruiters. One major research question, therefore, is whether the message content of military advertising in fact affects propensity and whether a move toward more values-oriented advertising would increase it. The other question is whether a move toward values-oriented advertising is more efficiently accomplished in a joint program or whether redirected Service programs would have the same impact at the same cost.

An advantage of loading all advertising into Service advertising was alluded to above: namely, that advertising unit costs are lower in programs of larger scale. This advantage probably disappears once programs reach some minimal scale. In fact, Navy and Army TV advertising seem to generate about the same impressions per dollar of spending despite the much larger scale of the Army TV program.[7] A potential advantage of having both joint and Service-specific advertising is that there are more competition and more innovation in the development and delivery of advertising the more firms are involved in the program.[8]

---

[6]The DSB report is available in The Defense Science Board Task Force on Human Resources Strategy, Office of the Under Secretary of Defense for Acquisition, Technology, and Logistics, Washington DC 20301-3140, February 2000. The Quality of Life panel was commissioned by Secretary of Defense Donald Rumsfeld and was chaired by Admiral David Jeremiah (Ret). It did not publish a formal report, but an informal report and briefing of its findings and recommendations are available from the Undersecretary of Defense for Personnel and Research.

[7]Unpublished calculations using the data supplied by PEP Research.

[8]In fact, there is some evidence from weapons programs that the cost-reducing competition from a second supplier outweighs the economies of scale that are lost when procurement is split between two firms.

The Target Audience

Current Service advertising is directed at youth themselves. The committee's previous report expressed the belief that some advertising needs to be directed toward civilian influencers of youth: parents, school teachers and counselors, and others (for the role of family in influencing behavior, see, for example, Sewell and Hauser, 1972; Otto, 2000; Mortimer and Finch, 1996). Such advertising, if effective, would build up the support base from which youth make early career decisions upon leaving high school. Changing the attitudes of adult influencers would not happen overnight and would require a sustained effort over time. One purpose of targeting adult influencers would be to counteract the decline in the adult veteran population, which one study (Warner et al., 2001) found to be related to the decline in enlistment in the 1990s.

An advertising campaign aimed at adult influencers would be consistent with the DSB recommendation to engage the American public about the value of public service.[9] It is an empirical question whether such advertising, if done properly, would have more impact on youth propensity—and the ultimate goal, enlistment—than advertising directed at the youth themselves.

Advertising aimed at adult influencers is likely to be more effectively accomplished in the context of a generic (joint) program than via Service-specific advertising whose purpose is to channel recruits to particular services. But that is an empirically testable hypothesis.

The Media

The Services advertise in a number of media: TV, radio, magazines, newspapers, direct mail, billboards, etc. In recent years about 60 percent of Army and Navy advertising expenditures have been for TV. Until recently, most Marine Corps advertising was TV and most Air Force advertising was non-TV. According to the Dertouzos-Garber model, an organization with a low advertising budget should spend the bulk of its budget on non-TV media. Non-TV advertising generates many more impressions per dollar than TV advertising (according to PEP data) and (theoretically) has a lower MES. But, theoretically, such advertising reaches its saturation point at a relatively low level of expenditure. TV advertising becomes cost-effective only at higher budget levels due to its higher MES, but it remains cost-effective much longer due to the fact that the saturation point is not reached until a fairly high level of expenditure.

[9] The committee's previous report also expressed the belief that more advertising should be values-oriented (National Research Council, 2003, p. 233-234).

These considerations suggest that the decision to eliminate joint TV advertising during the latter half of the 1990s as the joint program was scaled back made sense. However, devoting almost the whole 1989 joint budget of $20 million to TV advertising probably did not make sense.

Factors that would have to be considered upon any rebuilding of the joint program, or during experiments with advertising, are the message content and the target audience. Ads that appeal to patriotism, for example, hope to evoke a strong emotional response from viewers. TV or radio ads may be more likely to evoke such responses than magazine or newspaper ads. Messages that are targeted to youth in one medium may be more productively targeted to adults via other media. For example, if adults are more likely to read newspapers and less likely to watch MTV, ads aimed at adult influencers might be more effective in newspapers than on MTV.

Overall Scale of Recruiting

For several reasons alluded to above, the optimal levels of joint and Service-specific advertising are likely to depend on the overall scale of the advertising effort. In particular, joint advertising is likely to be more cost-effective in the context of a large recruiting effort than a small recruiting effort. This section discusses another reason why this might be the case.

Studies find a strong link between propensity to enlist, as measured from surveys such as the Youth Attitude Tracking Study, and actual enlistment (e.g., Warner et al., 2002). Given the youth propensity to enlist, the lower the military's demand for recruits, the greater the fraction of enlistments that will come from youth who are already positively inclined to enlist without military advertising to attract them. In such an environment, advertising can be targeted at those with a propensity and can be largely informational and aimed at informing them about career opportunities in particular Services and the like. It need not attempt to persuade youth that military service is a good thing; they are already so inclined.

But the larger the demand for recruits, the more effort the military must make to attract youth with no propensity to enlist. The issue is not "high" versus "low" demand on some absolute scale, but demand relative to propensity. Holding constant the recruiting goal, any change in propensity arising from factors other than military advertising (e.g., a lower unemployment rate) can influence the degree to which advertising needs to build propensity in the youth population rather than channel those who have already decided to join to a specific Service.

How a change in the scale of the overall recruiting effort affects the levels of joint and Service advertising depends in part on whether the two sources of advertising differ in the degrees to which they build propensity. Much Service advertising is aimed at inducing youth already with a propensity to enlist to join a particular Service and may not affect overall propensity in the youth population. To the extent that joint advertising can affect the propensity base, it should play a larger role in the overall advertising program when the larger the recruiting mission is compared with the base of youth with a propensity to enlist. As an example, joint advertising might not have been very productive in 1992, a very easy recruiting year, but it could have been very productive in the late 1990s, when propensity was lower and recruiting was much more difficult.

CONCLUSIONS AND RECOMMENDATIONS

We have addressed the question of whether there is a minimum level of advertising necessary for a cost-effective recruiting program, even if that advertising is not necessary to achieve current enlistment contract goals. Historically, when the recruiting climate is good and recruits are plentiful, military planners tend to cut advertising budgets, thereby contributing to a reduction in awareness capital and propensity levels. This may possibly set up a boom or bust cycle, in which propensity falls, recruiting becomes more difficult, and advertising funds have to be restored. We presented a model that describes the conditions under which it would be cost-effective to advertise in the interests of future enlistment supply, and we reviewed research to date that speaks to the issue. While extant research suggests that advertising may have effects only for a short period of time, the data available to previous researchers are limited for several reasons. First, they do not permit examining both lagged effects and nonlinear effects within a time period. Second, they focus on advertising aimed at youth at the point of the enlistment decision and do not permit examining possible supplemental advertising approaches, such as those aimed a youth several years prior to an enlistment decision or those aimed at adult influencers, such as parents.

As a result, research to date does not permit a definitive answer to the question of the cost-effectiveness of advertising above and beyond that which is necessary to achieve current recruiting goals. We recommend a focused effort to maintain advertising data in a systematic way for purposes of estimating a supply curve that incorporates the potential for both time-lagged and nonlinear advertising effects. We further recommend a program of research, incorporating quasi-experimental methods, to examine advertising effects over an extended period of time.

We then turn to the optimal levels of joint and Service advertising. It is our opinion that certain types of advertising themes, such as generic themes designed to increase overall propensity, are best done as a joint program, while advertising themes featuring specific benefits of military service are best done in the Service programs. What we do not know is whether there is an optimal level of joint and Service advertising or, more specifically, what advertising fund level should be allocated to joint programs. We note that issues of scale play a role in addressing this question, as certain types of advertising (e.g., television) do not appear to have a constant effect across levels of expenditure. The larger the recruiting effort and the larger the budget, the greater the potential value of a multifaceted campaign, with some resources targeted toward providing information about specific Services to those already with a propensity to enlist and others targeted toward increasing propensity among those currently without it.

We recommend a program of research aimed at examining the effects and cost-effectiveness of information-oriented versus values-oriented advertising in joint and Service advertising programs.

7

Determining Optimal Types of Incentives

Over the years of the All-Volunteer Force, various incentives have been developed and offered to help strengthen and shape military enlistments. The effectiveness of these incentives has been addressed and demonstrated, using a variety of evaluation approaches. This chapter considers methodological issues in determining the optimal types of enlistment incentives for specific recruiting goals. We begin by reviewing different types of enlistment options. Next, we consider the various types of effects the incentives may produce, both intended and unintended, and the related methodological issues in assessing their impact. We continue with a brief review of the analysis and methods discussed in the earlier chapters, which is followed by a longer consideration of analytical issues applicable to each of the evaluation approaches. Last, drawing on each of these areas, we conclude with a discussion of matching potential incentives and their effects to the appropriate assessment goals and evaluation methods.

TYPES OF ENLISTMENT INCENTIVES

Enlistment options may be grouped in various ways. For example, they can be distinguished in economic terms by whether they provide a cash incentive, cash in kind, or some other feature valued by potential enlistees, such as conditions of service. An alternative is to group incentives according to the segment of the recruiting market they are intended to attract, such as youth interested in attending college or those looking for job training. In the discussion that follows we use a mix of these two approaches. We do so because that classification scheme lends itself directly to our discussion of the possible effects of these incentives, appropriate assessment goals, and evaluation methods.

Youth also enlist for intrinsic reasons, such as patriotism. Matching recruiting messages and strategies to such motivations was discussed earlier. In developing or assessing the effectiveness of such strategies, considerations analogous to those discussed for the options considered in this chapter apply. More generally, enlistment options can be categorized based on social-psychological or economic theory according to their intended effects on the determinants of enlistment behavior, and their success in achieving these effects can be assessed as part of the evaluation process for the options.

One group of incentives consists of cash enlistment bonuses. These bonuses are offered for various purposes: increasing enlistments; channeling recruits into hard-to-fill occupational specialties; encouraging recruits to report for active duty quickly; encouraging enlistments or accessions during the seasons of the year, such as late winter and early spring, that pose chronic challenges; encouraging enlistments among persons who served previously but left the military; and so forth.

Military pay increases also are used to increase enlistments. Such increases are less targeted or flexible than bonuses and are intended to provide an overall enlistment stimulus. Compensation increases also can be provided through changes in other benefits, such as housing or subsistence allowances.

Another group of options consists of incentives provided to encourage youth interested in postsecondary education to see military service as a facilitator of their educational aspirations, rather than as an alternative career option. (As discussed in Chapter 2, this amounts to changing their beliefs about the consequences of enlisting compared with options in the civilian sector, and it can also change the analogous views of their key influencers.) Such education incentives include credit for college courses completed prior to accession (such as credit toward meeting military training requirements); near-term financial incentives, such as entry at an advanced pay grade, student loan repayment, or a bonus; funding for college after service, for example, through the Montgomery GI Bill or the Service's college fund; money for attending college before service, through options such as the Army's College First program (currently being pilot tested); or in-service continuing education programs, in which classes are provided on site or through distance learning.

A fourth type of incentive is directed specifically at improving the recruit's job opportunities. Included in this group are such options as the

Army's Partnership for Youth Success (PaYS) program and the GED Plus pilot program.[1]

Another group of incentives involves choices concerning the conditions of one's military service. Included in this group are such options as the term of enlistment, location of assignment, and combination of active and reserve forces duty, among others.

Of course, the Services can and do use specific enlistment options in combination with others. One example is the combination of enlistment bonuses with postservice educational benefits. Another is the combination of assistance for attending college classes with financial compensation or training credit for having done so.

POTENTIAL EFFECTS OF INCENTIVE

To structure the analysis of the effects of a current or prospective enlistment option, it is important to first address several key issues. These issues include, but are not limited to, the types of questions highlighted below.

- Who is the target of the option? For example, is the option intended to increase overall enlistments or specifically those among highly qualified youth (high-aptitude high school graduates)? Similarly, are there key subgroups of interest in the target population? If the incentive is intended to attract college-bound youth into the military (by encouraging them to see military service as a facilitator of their educational aspirations, rather than as an alternative career option), is it aimed at college graduates, current students, college stopouts (those who attended some college, have not yet graduated, but are not currently enrolled), high school graduates who have not yet attended college, high school seniors interested in college, or some combination of these subgroups?
- What are the desired effects of the option? These can include such outcomes as increased supply through increased enlistments or market expansion (or improved recruit performance through increased supply among high-aptitude youth), skill channeling into hard-to-fill or new spe-

[1]As is true for postsecondary educational incentives, the PaYS program is intended to change youths' beliefs about the consequences of enlisting compared with options in the civilian sector—in this case by providing job training that can be applied to the civilian sector and helping the youth to be considered for employment in a related job after service—and it can also change the analogous views of their key influencers. Among its other goals, such as increasing training opportunities, GED Plus is designed to facilitate enlistment among high school dropouts with positive propensity while they achieve the qualifications needed to serve on active duty.

cialties, increased man-years, increased affiliation with the reserve forces following active duty, cost savings, and so forth, singly or in combination.

- What are possible unintended or undesirable effects of the option (for example, on such matters as the quality of enlistments, term of service distribution, attrition, or retention)?
- What are the potential timing and magnitudes of the desired and unintended effects?
- What are the option's potential costs? This includes such considerations as the direct cost for new recruits, overall cost including economic rent, (discounted) future costs, training costs, costs or savings to the reserve forces?
- How much do we know about the types and specific features of incentives that might generate effects analogous to those desired from the option? Does the option differ in important ways from past incentives for which we have enlistment data?

We consider each of these issues in turn.

Who is the target of the option? Enlistment incentives are typically targeted according to the qualifications of the recruit and the occupational specialty in which he or she enlists. Qualifications often refer to whether or not the recruit has a traditional high school diploma and the level of his or her (written) aptitude score on the Armed Forces Qualification Test (AFQT). Youth with diplomas who score in the upper 50 percentile on the AFQT are considered to be highly qualified and are eligible for the largest incentives (subject to meeting the incentives' other requirements). Alternatively, an option such as GED Plus may be targeted to the non-high school graduate market.

Within the target population as a whole, there may be subgroups of special interest. This heightened interest may reflect a Service's desire to improve its recruiting performance for a particular youth subgroup. For example, in the case of GED Plus, one such subgroup consists of Hispanic youth. Alternatively, there may be heightened interest in specific subgroups because the subgroups require different marketing or recruiting activities to effectively employ the option. In the case of incentives designed to appeal to college-bound youth, as noted, such subgroups may consist of high school seniors, high school graduates without college, college stopouts, youth currently enrolled in college, and college degree holders. Another relevant consideration concerns the option's potential appeal to the targeted youths' key influencers or referents (see Chapter 2).

What are the desired effects of the option? The incentive can be intended to affect a number of domains. These include, but are not limited to, increasing the enlistment rate among subgroups of the youth population who currently are enlisting; expanding enlistments to other subgroups; channeling enlistees into specific military occupational specialties; increasing the number of years served by the typical enlistee; increasing affiliation with the reserve forces following active duty; or reducing recruiting costs. Defining the desired effects of the option is central to defining the outcome measures needed to assess its effectiveness.

What are possible unintended or undesirable effects of the option (on such matters as the quality of enlistments, the term-of-service distribution, attrition, or retention)? In addition to its desired effects, the option may have unintended consequences, which may be significant and thus require assessment. For example, an incentive designed to increase enlistments among high-aptitude GED holders should not do so at the expense of enlistments among high-aptitude high school graduates, which is the primary recruiting market. Alternatively, youth joining the military under an incentive designed to expand the recruiting market by offering shorter terms of service should not be those who would have enlisted for longer terms under existing programs. When incentives have potentially significant adverse outcomes, the unintended effects should be assessed in the evaluation together with those desired. Options may also have more subtle unintended consequences that nonetheless are very important to address in assessing the options' efficacy. For example, to the extent that the incentive increases youth-propensity to enlist for a particular market or subgroup, recruiters may decrease the effort they expend for that market or for other subgroups (increase environmental constraints). This is because recruiters may not need to leverage the full potential of the incentive in order to meet their recruiting goals and thus can target their efforts. This can lead to underestimating the option's actual potential to increase enlistments or expand the market, and it needs to be addressed specifically in the analysis.

What are the potential timing and magnitudes of the desired and unintended effects? The timing of the option's effects has a direct bearing on the period of the evaluation. Options designed to increase accessions into the military in the near term may require a relatively short assessment period, whereas options designed to increase man-years or reserve forces affiliation may require that enlistees under the option be followed up through their first term of service (or even longer). Still other options designed to increase accessions may do so only over a longer period of time. Incentives provided to allow youth to attend college classes following enlistment but prior to entering active duty fall into this category and thus require a longer evaluation period to cover the increased time between the enlistment and accession points. A longer period between enlistment and accession would be expected to lead to higher loss rates among enlistees before their reporting for active duty. As a result, capturing such loss rates becomes an important component of the analysis. Options that increase enlistments in the near term may also have longer term consequences. Postservice educational benefits might be expected to increase enlistments but also to increase separations if the recipients leave the military at a higher than average rate in order to attend college.

The anticipated magnitude of the incentive's effects has a direct bearing on the statistical power of the analysis. Consequently, it affects the number of observations required to make a reasonable assessment of the viability of the option. This, in turn, can affect the number of persons to whom the option is offered as well as the required duration of the assessment.

What are the option's potential costs (such as the direct cost for new recruits, overall cost including rent, discounted future costs, participation rates, training costs, reserve forces costs or savings)? Enlistment options that have relatively small elasticities in increasing enlistments can nonetheless be very cost-effective recruiting tools depending on their cost. Conversely, the utility of incentives with high elasticities can be limited by their cost. To the extent that the costs of an enlistment incentive are of interest, they need to be captured in the evaluation. This can include the costs (and effects) of alternative values of the incentive or of variations in the number of specialties for which it is offered. The cost of providing the option to qualifying enlistees typically also includes costs in the form of "economic rent." This applies to the payment of the incentive to recruits who would have enlisted even in its absence (a cost we seek to minimize relative to the cost incurred to increase enlisted supply).[2] A related cost concerns the impact the option may have on enlistments under other incentives and the resulting overall net cost increase or savings.

Incentives that are paid out only after a considerable period of time, such as postservice educational benefits, have true costs well below their nominal value, unlike enlistment bonuses. Thus, their cost needs to be discounted in light of the delay in providing them, and the eventual usage rates for the incentive need to be considered. This can include both the proportion of eligible recipients who use any part of the incentive as well as the proportion of their entitlement they eventually use. Enlistment incentives can also increase or reduce other types of military costs. For example, an enlistment option combining active and reserve forces duty can reduce reserve forces recruiting costs by increasing affiliation rates

[2]The benefits considered in the rent calculation can include more general benefits to society provided by the incentive, in addition to increased enlisted supply. An incentive that increased college attendance or provided civilian-related job training could be expected to improve the employment opportunities and earnings of the recipients (and the taxes paid by them) over the longer term.

from the active force into the reserve forces and training costs by encouraging affiliation in the occupational specialty the soldier held while on active duty. Further, enlistment options can affect attrition and retention rates, which have implications for recruiting and training costs.

How much is known about the types and specific features of incentives that may generate effects analogous to those desired from the option? Does the option differ in important ways from past incentives for which we have enlistment data? The extent to which a prospective option differs from current and past options has a direct bearing on the requirements for new data collection to assess its effectiveness. At one extreme, consider a new enlistment bonus whose value lies between the values of enlistment bonuses already offered. In such a case, it is very likely that existing data could be used to estimate the effects of offering the new incentive. Alternatively, consider the case in which a prospective enlistment option differs in important ways from those currently available in its basic features or in the portion of the youth market to which it is targeted. An example would be a program with unique features, such as PaYS, or a program designed to attract youth who would not otherwise have joined the military, for example, by providing programs to help them attend college between the enlistment and accession points.

EVALUATION METHODS AND ISSUES IN THEIR USE

Four major types of evaluation approaches are commonly used in the development and assessment of recruiting incentives:

- Focus groups,
- Surveys,
- Econometric (regression) analyses of existing data, and
- Experimental or quasi-experimental assessments.

Each of these approaches has advantages and limitations. Focus groups lend themselves especially well to exploring in depth the attractive and unattractive features of a limited number of prospective or current incentives (or to helping develop new types of options), whereas surveys can be especially useful in more general assessments of the relative attractiveness of a set of specific alternative options to the recruiting population as a whole or among particular market segments. Regression analyses are especially helpful in assessing enlistment rates under current (or past) options and in understanding the possible effects of trade-offs among the options or of limited adjustments to the options. When we need to estimate enlistment rates under prospective options that differ significantly in their features or market segment from current and past options, then experimental or quasi-experimental designs can be particularly useful.

We next consider the strengths and weaknesses of each approach in detail. We note here that each approach can have both unique advantages and limitations and this often leads to the use of multiple assessment methods in the evaluation of recruiting options.

Focus Groups

The use of focus groups is especially helpful for shaping prospective recruiting programs by enabling the researcher and policy maker to explore the appeal of a range of incentives or enlistment options to youth or their influencers, including programs currently available to recruits as well as prospective options. Moreover, the focus groups can provide insight into the particular strengths and weaknesses of such programs, including the usefulness and appeal of specific features of the options and the extent to which that appeal varies across key subgroups of the recruiting market. In so doing, the focus groups can facilitate the discovery and exploration of new option features or effects that may be helpful in achieving military recruiting goals. A further advantage of this approach is that the information can be gathered relatively quickly and inexpensively.

The use of focus groups is not without limitations. The groups must be carefully constructed to represent the population of interest and to facilitate open discussion. Groups can be too large to generate open discussion, can be affected by social desirability concerns or other response biases, or can be dominated by a small number of especially vocal participants (or by the group leader). Even if these potential obstacles are overcome, the groups still have only a limited ability to provide information that is generalizable to the full population of interest, because of the limited number of participants and the semistructured nature of the discussions. This is particularly true when the purpose of the evaluation is to estimate effects of prospective alternative incentives or options such as the preference rates for one option over another or actual enlistment rates.

Surveys

Surveys also can provide rapid and reasonably inexpensive assessments of existing or prospective enlistment options among youth or their influencers. Because they are administered individually and, moreover, the questions can vary across individual respondents, they can be constructed to cover more ground than what can normally be addressed using focus groups. This is particularly true for surveys administered using computers, such as computer-assisted telephone interviews or web-based surveys; in such surveys, an individual respondent's answers to specific questions as he or she proceeds through the survey can be used to determine which succeeding questions he or she is asked. For example, surveys are ideal to assess potential interest of youth in military service; can be designed to facilitate comparisons of the attractiveness and effects (among youth or their influencers) of a substantial number of specified, alternative recruiting options; can be designed to readily assess the variation in propensity and in the effects of such incentives for key youth population subgroups of interest; and can be designed to explore the covariation of enlistment interest and the appeal of various incentives (or their specific features) with a wide variety of other factors, such as demographic factors, educational aspirations, job and career goals, reasons for enlisting and barriers to military service, discussions with influencers, contacts with the military, and awareness of current military benefits. Questions dealing with such matters as media preferences, Internet use, and recreational habits can also be included to provide insights into the most effective ways to market enlistment incentives, both overall and to specific subgroups.

As is true for focus groups, there are important issues and limitations in the use of surveys to assess the effects of alternative enlistment incentives on military recruiting. Drawing firm conclusions regarding the efficacy of alternative enlistment options requires that the sample must be representative of the target population, that the sample size must be adequate, and that a sufficient proportion of the surveys are completed and returned. Determining the appropriate sample size involves decisions about the acceptable confidence intervals surrounding the key point estimates and differences (in responses to specific items) to be provided by the survey; the statistical variances of the corresponding items; the fraction of respondents that will be asked to answer the key questions; the estimated completion rate for the survey; and the acceptable level for the statistical power of the analysis to detect true differences.

Even when the sample is adequate and representative, care still must be taken in interpreting survey results concerning the prospective effects of alternative enlistment options. Although statistically significant relationships between one's stated likelihood of enlisting in such surveys as the Youth Attitude Tracking Study (YATS) and Monitoring the Future (MTF) and one's actual enlistment decision have been demonstrated, it also is true that respondents' misestimate their true enlistment probability (Bachman, Segal, Freedman-Doan, and O'Malley, 1998; Orvis, Gahart, and Ludwig, 1992). Youth seeing themselves as very likely to enlist are, in fact, less likely to actually join the military than they believe, whereas youth seeing themselves as very unlikely to enlist are more likely to do so than they believe. The accuracy of a respondent's stated enlistment intentions varies with a number of factors, such as the temporal proximity of the enlistment decision point and the underlying rate of enlistment in the respondent's demographic group (normative support). When there is a considerable time period between the intention assessment and actual enlistment decision point, for example, changes in the respondent's life, in world events, in the economy, or in the information obtained through discussions with key influencers can alter the youth's enlistment intentions. This also limits the ability of surveys of the youth population to capture the possible longer term costs and effects of alternative enlistment policies, such as the incentive's effect on attrition and retention, even if the survey includes questions concerning such effects.

Past research also suggests that youth overestimate their true responsiveness to changes in enlistment options. While surveys provide an excellent methodology to assess the *relative* effects of alternative enlistment incentives, obtaining good estimates of the true enlistment rates under the options is likely to require additional data. Moreover, the specific details of the information provided to respondents concerning current or prospective enlistment options are important and can influence their answers; this involves matters both of question wording and order, as well as social desirability concerns and other response biases. Finally, surveys of enlistment interest in the youth population (the supply side) do not capture the potential behavior of recruiters and recruiting managers as the recruiting market changes or new incentives are introduced (environmental facilitators or constraints—the demand side); this again limits the ability of survey data to provide accurate estimates of the true enlistment rates that would be obtained under alternative options.

Econometric Models

As discussed in some detail in Chapter 5, econometric approaches rely on existing data and thus can provide rapid and inexpensive assessments and comparisons of policy alternatives. They also can provide statistical controls for confounding factors—such as historical differences in enlistments across recruiting areas or the effects of changes in the economy—and can allow assessment of incentive effects for subgroups of the youth population. Econometric models can assess the effects and costs of enlistment options over time and in different recruiting environments. They provide data on actual enlistment rates and can directly provide data on the magnitudes of unintended effects; such effects can include market substitution—the substitution of enlistees brought in under a program for other types of enlistees or for enlistments under other options that would have occurred in the absence of the program—and can be used to correct for changes in demand-side behaviors (environmental constraints or facilitators) that may mask the true potential of an option to increase enlistments or expand the recruiting market (Buddin, 1991; Polich, Dertouzos, and Press, 1986).

Although their reliance on existing data offers advantages of speed and cost in assessing the efficacy of alternative recruiting options, it limits the usefulness of econometric models in shaping and assessing prospective options. Because the models are limited to assessment of previous or current recruiting incentives, the possible effects of new options that differ in important ways from such incentives—in their features or the market segments to which they are intended to appeal—cannot readily be assessed. Moreover, the accuracy of econometric models in forecasting enlistment rates can be limited by a variety of other factors that merit careful attention by the analyst. These include, but are not restricted to, the often limited amount of enlistment variation they may explain, especially at the individual (versus cross-sectional) level or over time; the possible misspecification of the model (for example, due to the inclusion of an insufficient set of predictors or use of an inappropriate functional form); failure to capture important recruiter mission-setting or management practices that affect enlistments; failure to capture major shifts in propensity that may occur over time; and the possible correlation between supply- and demand-side variables in the model, such as changes in resources in response to changes in enlistment propensity. If recruiting resources are increased in response to a decline in youth's propensity to enlist and it then takes a while for the resources to improve the recruiting environment, one could underestimate the efficacy of the resources or, in the extreme, obtain incorrect signs for their regression coefficients. Similarly, if recruiters reduce their effort in response to the efficacy of an incentive in increasing enlistment propensity, the potential effects of the option on increasing enlistments or creating market expansion can be underestimated.

Experimental and Quasi-Experimental Designs

Use of experimental or quasi-experimental designs to evaluate the potential effects of enlistment options offers several advantages. The recruiting effects of prospective enlistment incentives can be quantified at low risk compared with full implementation (e.g., by avoiding unnecessary costs or the difficulties of subsequently withdrawing an established, national incentive). Moreover, the effects of alternative prospective policies or of alternative versions of a prospective policy can be directly assessed and compared. The effects of potential confounding factors in assessing the efficacy of recruiting programs—which need to be statistically controlled after the fact in econometric models—can be directly and greatly reduced through balancing of the test areas offering new programs and the control areas not offering the programs on factors that are known to affect enlistment rates. These include, for example, geographical location, local economic conditions, demographic characteristics, past recruiting production, local recruiting missions, such recruiting resources as the number of recruiters in the area, and the alternative enlistment options that may be available locally, particularly to the extent that other Services or reserve components offer related programs (for example, alternative financial assistance programs for college attendance).

By employing an experimental design that compares (1) changes in enlistments in the test areas under the new option relative to enlistments in the same areas during a pretest baseline period with (2) the trend in enlistments in the control areas not offering the new program over the analogous time periods, the approach provides a direct control for temporal and fixed differences in enlistment rates. Experimental and quasi-experimental designs can also directly provide data on the magnitudes of unintended effects; as noted, such effects can include market substitution or changes in recruiter effort or time allocation in response to the change in enlisted supply created by the new incentive (such as reduction of effort as supply increases).

Given a test period of sufficient duration, the experiment's interim results, concerning both desired and unintended effects, can be used to recommend adoption of the test program or to provide the basis for changes in the incentive or in recruiter mission for the incentive that are designed to enhance its efficacy. The remainder of the test period can then be used to evaluate the performance of the modified program.

Use of experimental or quasi-experimental designs to test the potential effects of alternative options is not without potential problems. Certain recruiting incentives, such as military pay increases, simply do not lend themselves to pilot testing; it is not practicable to limit them to certain areas of the country or to undo them if their effect is not as desired. In cases in which legal authority is required to test prospective options, obtaining such authority may require considerable time, and the specifics of the enabling legislation can compromise the test's results. Similarly, the specific details of the incentive's test implementation—such as the proportion of occupational specialties eligible for the incentive—have a direct bearing on the estimates of the enlistment effects of the incentive that will be derived from the experiment.

Such issues require careful attention in the analysis, particularly if comparisons with other incentives that differ in such characteristics are desired (see, for example, Buddin and Roan, 1994). Additional resources for the test programs may also be needed, particularly if the experiment is to continue over an extended time period. A related issue concerns the feasibility of conducting a stable test in the operational environment, particularly if recruiting conditions improve or worsen significantly during the test period or if the experiment suggests the usefulness of the pilot program and other recruiting areas (or Services) decide they want it right away.

Moreover, there are certain inherent limits when experimental or quasi-experimental designs are used to assess the effects of prospective enlistment options. These include the fact that the program attributes that can be evaluated often are more aggregated or limited in number than is the case for some of the other evaluation methods; this can occur due to resource constraints, policy-based limits on enlistments under the pilot program, or statistical requirements to support the reliability of the test results, among other factors. Also, when new programs are tested in an experiment—as compared with econometric analyses of alternative options using existing data, for example—their overall cost-effectiveness may be unknown for years; this is because such a determination requires assessment not only of enlistments—which can be measured in the relative near term experimentally—but also of patterns of attrition, retention, and job performance, which take much longer to unfold. Finally, no matter how well balanced the test and control cells may be at the outset of an experimental assessment, the test does not control for differing intertemporal changes across the cells, such as those that could occur due to changes in local economic conditions or changes in demand-related factors, such as recruiting behavior or resourcing.

Combinations of Evaluation Methods

Sometimes a combination of evaluation methods is used to assess the efficacy of alternative enlistment options. One approach that has been used in the past is to combine econometric analysis with the experimental approach in order to help preserve the balance of the test and control cells by adjusting statistically for unexpected, differing intertemporal changes across the cells.

The other methodologies also can be used in combination. For example, focus groups may be used to help explore and define prospective enlistment options that are then tested in large youth market surveys or in pilot tests employing experimental designs (or, conversely, focus groups or surveys are used to explore the effects of policy experiments). Another example is the application of econometric methods or other analytical techniques to existing datasets in order to help interpret survey results. The other approaches could be used to derive a metric to help translate stated enlistment intentions in the survey into estimated enlistment rates.

ASSESSMENT OF INCENTIVE EFFECTS

In the final section of this chapter, we provide examples of designing appropriate research to help determine the optimum type(s) of enlistment incentives for a given policy goal by matching the desired effects of the potential incentive(s), the corresponding assessment goals, and evaluation methods.

Consider the case in which the Office of the Secretary of Defense (OSD) or an individual Service would like to expand its recruiting efforts into a new or largely untapped market and is interested in developing enlistment incentives to do so. To the extent that the market is new or exploratory research on new options is desired, econometric modeling using existing enlistment data will not be especially helpful in choosing optimal incentives for recruiting this market. Experimental or quasi-experimental approaches would limit the number of options that could be tested, due to analytical, cost, and legislative considerations, among other factors, and thus are premature. The policy maker also lacks sufficient information at this point to design a survey to gather detailed information on the areas of interest. Focus groups, however, are well suited for this purpose.

The Army's interest in materially improving its recruiting performance among youth who are committed to attending college after high school but would consider enlistment subsequently provides an example. Over the past 20 years, the proportion of youth attending college within a year of high school graduation has risen dramatically, from about half to approximately two-thirds (U.S. Bureau of the Census, 2000, 2002). However, the military has lagged in adjusting its recruiting strategy and still recruits the vast majority of its enlisted force from youth who have not attended college. In recognition of this mismatch, over the past several years the Army has employed policies that give greater emphasis to recruiting youth from the college market (which includes students, stopouts, and high school seniors). As part of that effort, it has conducted focus groups with youth in the college market to better understand their reasons for or against enlisting and the types of enlistment options they would find appealing. Similarly, OSD recently sponsored a study of the college-bound youth population concerning their enlistment interest and the types of incentives that would stimulate that interest (Kilburn and Asch, 2003). The initial step in the research consisted of a series of focus groups with each of the subgroups listed above; this helped OSD define and shape incentives that appeared to be promising for further investigation.

Next, consider the case in which a policy maker wishes to understand in some detail the relative appeal of an array of alternative recruiting options and wants to be able to generalize the information to the youth population. The options may include current incentives, prospective enlistment incentives, or both. The details of interest, for example, may include the usefulness of the incentives for key youth subgroups, such as high aptitude youth and college stopouts; those potentially interested in military service for various reasons, such as job training, assistance for attending school or taking a break from work or school; or those with specific barriers to military service. If the policy maker wants to compare current and prospective incentives, it may also be helpful to assess awareness levels in the subgroups of interest for the significant current options.

To the extent that research on new options is desired or the youth population characteristics of interest are not captured in the military's enlistment databases—for example, starting and then stopping out of college before completing a year, recruits' reasons for enlisting, the level of support for enlisting among their influencers, or youths' awareness levels of specific options—existing enlistment data will not be especially helpful in choosing optimal incentives. Experimental or quasi-experimental approaches would again limit the number of options that could be tested, and they might not identify key subgroups of interest without additional data collection procedures. Given the policy maker's interest in systematic information that can be generalized to the youth population, focus groups lack the required coverage and rigor. In such instances, youth surveys can be used very effectively to address the issues of concern. In the OSD study just discussed, the incentives developed using focus groups were included in a subsequent survey of college-bound youth that provided a variety of detailed, generalizable enlistment-related information, including the incentives' potential effects on enlisted supply. Similar considerations apply to the use of influencer focus groups and surveys.

As discussed earlier in this chapter, surveys must sample and obtain data from a sufficient number of respondents representative of the population and subgroups of interest. Care also must be taken in the wording and sequencing of the questions included to represent the enlistment options. The results will be best suited to understanding the relative effects of the alternative incentives. Using propensity measures for which data exist to link responses to enlistments will help to meaningfully interpret the responses to the options in the survey by providing an enlistment index value for each option and subgroup. Even so, past research has shown that respondents misestimate their true enlistment probability to some extent, and there also is the possibility of changes in demand-side factors, such as recruiter behavior in response to new markets or options. Thus, the index values do not equate to the actual enlistment rates that will be obtained under the options. There are various analytical approaches that can be used to construct the samples, survey questions, and enlistment indexes (see, for example, the Youth Attitude Tracking Study, 1976-

1999—Wilson et al., 2000; Kilburn and Asch, 2003; Orvis and McDonald, forthcoming).[3]

A policy maker's interest in determining the optimal types of enlistment incentives may involve wanting to optimize the cost-effectiveness of the mix of current options or to get the greatest return on investment for additional recruiting resources that can be made available in the near term. By analogy, the issue could involve wanting to reduce a particular type of resource—for example, recruiters—and wanting to know how to best offset that reduction with alternative resources. Answering these types of questions quickly and with confidence requires estimates of actual enlistment rates, costs, and elasticities, based on known past results. Econometric modeling is ideal in such cases (see, for example, Warner, Simon, and Payne, 2001; Buddin and Roan, 1994; Asch and Orvis, 1994). Using econometric modeling to address such questions also has the advantage of enabling the policy maker to examine in advance the probable longer term consequences of making such allocations and trade-offs on such factors as attrition and retention.

---

[3]The purpose of an enlistment index is to allow the prospective recruiting effects of alternative options to be compared more meaningfully than would be possible simply by comparing the distributions across an intention measure generated by the options. There are various approaches to constructing such enlistment indexes. They all include linking enlistment records to stated enlistment intention levels—normally, from earlier surveys—in order to establish enlistment rates for the intention levels. For example, this can be done for the individual levels (response categories) of the intention measure (e.g., definitely, probably, probably not, definitely not or 0 to 10 likelihood of enlisting); for combinations of the categories (e.g., positive versus negative propensity); or for composite measures that combine more than one intention measure (e.g., "definitely" on any of several individual Service measures becomes "definitely" on the composite measure, or an unaided mention of plans to enlist forms the top category, followed by positive propensity without an unaided mention, and so forth). Various follow-up periods can be used to determine the enlistment rates—such as one year, lifetime, etc.—depending on one's purpose (and the ages or school year of the respondents). It is important to match the intention measures to be used in the new survey, the sample composition (e.g., market segments, rules on inclusion of youth in the delayed entry pool), and the enlistment period of interest to those used in the earlier follow-up work to generate the index. The enlistment index value can simply be the enlistment rate found in the enlistment records check for respondents stating the given intention level, or it can be transformed or made conditional on other factors (e.g., through inclusion in a regression analysis of enlistments) to help control for the effects of other factors on a person's enlistment decision, which may not be captured fully by the intention measure. It may also be helpful to use the index values to compute a measure of the options' appeal relative to each other, by scaling the index values relative to the most (or least) appealing option (i.e., that which most (or least) increases stated propensity to enlist in the survey). Finally, the index values may be calculated independently or relative to that for the respondent's (baseline) likelihood of enlisting under current options.

The data collected from focus groups is not sufficiently generalizable or precise to address these types of questions. Although surveys can be useful in comparing the relative appeal of alternative options, they do not provide absolute enlistment (or loss) rates, and they do not directly capture the effects of changes in such factors as the number of recruiters or advertising levels. Finally, given the focus of these questions on trade-offs among existing incentives and resources, using an experimental approach would not provide results in a timely manner and would be unnecessarily restrictive.

When the types of incentives to be evaluated are reasonably well defined but they are new in total or vary significantly in their features from current options, assessing their effects requires new data collection. Collecting that information with reasonable precision—beyond that afforded by surveys or focus groups—but with lower risk than through national implementation can be accomplished using experimental or quasi-experimental pilot tests of the programs. The general features of such tests, as well as their strengths and limitations, were described earlier. There are numerous examples of national recruiting tests, and they cover a variety of effects of potential interest to policy makers.

For example, the Educational Assistance Test Program compared recruiting effects for the Army, the Navy, and the Air Force of different types and amounts of postservice educational benefits (Fernandez, 1982). Issues of concern included the overall effect on highly qualified enlistments of the incentives; the effects of the type and value of the incentive; and the effect of differences in the value of the incentive across the Services. The Enlistment Bonus Experiment compared the results of bonuses differing in dollar value and term of enlistment obligations on highly qualified enlistments and the channeling of recruits into hard-to-fill occupational specialties (Polich et al., 1986). The 2+2+4 Recruiting Experiment addressed the effects on active and reserve forces component enlistments among highly qualified youth of a program allowing two years of active duty plus training time followed by two years in the Selective (drilling) reserve forces and four years in the Individual Ready Reserve in return for enhanced postservice educational benefits (Buddin, 1991). The key effects of interest included active enlistments, changes in job choice and term-of-service obligation, affiliation rates into the Selective Reserve, affiliation rates into the same occupation held while on active duty, and the overall impact on military man-years. The 2+2+4 test used a true experimental design that included random assignment of qualifying youth to eligibility for the program.

Another example of employing a quasi-experimental design is provided by the College First/GED Plus National Recruiting Test. These pilot programs are currently in progress (Sellman, 1999; Orvis and

McDonald, forthcoming). The College First program's main purpose is to assess the effect on highly qualified enlistments and market expansion of an incentive allowing recruits to attend college for up to two years after enlisting but before reporting for active duty. The analysis examines such effects both overall and for key subgroups of the college-bound youth market. Based on these analyses, it identifies useful changes in the program and in recruiter missioning to support it. GED Plus was designed to increase opportunities for and enlistments among qualifying youth without traditional high school diplomas. To qualify, youth have to score in the upper half of the AFQT distribution and pass strict screening criteria. Youth without GED certificates have to obtain them through an attendance-based course prior to going onto active duty. Key outcome measures include the program's effects on high-aptitude enlistments among youth without traditional high school diplomas and, in particular, among Hispanic youth, who are disproportionately represented in this market; the impact, if any, on highly qualified enlistments (market substitution); first-term attrition rates under the program; and the cost-effectiveness of eligibility for alternative enlistment incentive levels among GED Plus recruits, including the extent to which each level is being leveraged by recruiters.

Helping policy makers choose the optimal incentives to accomplish their objectives and obtaining reasonable estimates of the incentives' probable effects often involves the use of multiple evaluation methodologies. We have noted such combinations in the preceding discussion. For example, focus groups can provide a highly useful initial step in designing new enlistment options. The main results of the groups can then be used for follow-up assessments that provide more rigorous and generalizable results through surveys or recruiting experiments. We also noted that quantitative assessments linking prior survey results to actual enlistment data can be very helpful in interpreting enlistment propensity information collected in new surveys and in quantifying the likely enlistment outcomes associated with alternative incentives.

As discussed earlier in this volume, econometric modeling can be used in conjunction with experimental or quasi-experimental designs to provide improved estimates of the potential recruiting benefits of new options. The Enlistment Bonus Experiment, for example, used econometric analysis to support a quasi-experimental design. The regressions controlled for changes over time in economic and other conditions that could have acted to unbalance the test's experimental and control cells. Based on Dertouzos' approach, they also attempted to control for unobserved reductions in recruiters' effort in response to the improvement in recruiting conditions under the test options and the ease of substitution of less-well-qualified versus highly qualified recruits.

For illustrative purposes, we conclude with an example of using the four evaluation methodologies in combination. The Army is currently assessing the benefits of eArmyU, an online distance learning program allowing soldiers to pursue courses and degrees from more than 20 colleges and universities and approximately 100 programs, and of soldiers' interest in modified versions of the program. The current program provides a free laptop computer to participants, up to 100 percent tuition assistance, and a number of other benefits, such as an Internet service provider and free books and delivery. In return, the soldier must agree to serve for at least three years from the eArmyU enrollment point and must complete 12 semester hours within two years of enrolling.

The evaluation is assessing participation rates in the current program across subgroups (using econometric modeling); retention effects of the current program (econometric modeling and focus groups with participants in the program); and likely enrollment rates and retention effects under alternative versions of the program—some with reduced incentives matched by reduced service and course completion requirements, in order to allow more soldiers to participate within the resources available for the program (quasi-experimental pilot test, survey of enlisted personnel, and focus groups with participants, supervisors, and education administrators of the alternative versions of the program). The evaluation also is assessing quality of life benefits of the current program to the soldier and his or her family (focus groups) as well as the effects of the current program on readiness and duty performance (focus groups).

8

Performance Management of Recruiters

Earlier chapters described various approaches that can be used in an evaluation framework for decision making. Econometric, experimental, qualitative, and survey approaches all have their place in this framework. We have seen how each of these approaches has strengths and weaknesses that play a role in determining which approach (or approaches) to use in a particular situation. Many of our examples have focused on recruiting resources and how best to use them. For example, we have addressed evaluation of advertising themes, the appropriate levels of joint and Service-specific advertising, various incentives (and the levels of incentives needed) to attract recruits, and others. In this chapter, we focus on what Barnes, Dempsey, Knapp, Lerro, and Schroyer (1991) refer to as "the linchpin to recruiting success"—the recruiter. Recruiter performance management encompasses the range of issues and decisions that face Service recruiting managers as they organize to meet their mission. We demonstrate that effective performance management requires multiple evaluation methods.

ISSUES IN PERFORMANCE MANAGEMENT

Service recruiting managers establish systems to select recruiters from among the available pool of Service members, to train and develop those new recruiters, to open recruiting offices in specific locations, to establish production goals for each recruiter, to motivate recruiters with reward and recognition programs, and to monitor and assess recruiter performance. Many models and options are available for each of these systems, and each model or option chosen is open to evaluation. In some cases (for example, selection of new recruiters), there are continuing research programs to evaluate the effect of alternative programs. In other cases (for example, the effects of recruiter reward and recognition programs), research or evaluation is rarely attempted. In still other cases (for example, establishing recruiter goals), there is a research base for some aspects of the program (for example, determining market size) but very limited research on the effectiveness of alternative approaches to actually setting goals. Given the central role that recruiter productivity plays in the recruiting process, all aspects of recruiter performance management should be subjected to evaluation efforts.

Performance management systems are certainly affected by the environment in which the recruiter works. (By environment, we mean all of the conditions that surround recruiters performing their jobs—including the environmental conditions created by the military and those that are broader and culturally based.) As stated in the committee's earlier report, the achievement of recruiting goals can be highly dependent on the economic conditions of the time, with high unemployment rates resulting in the easier attainment of recruiting goals. Moreover, other environmental factors, such as well-publicized military actions, proximity to military bases, and recruit's acquaintance with soldiers, may have some impact on a recruiter's performance. Despite the importance of these environmental factors, we have not addressed them specifically. In many cases, we recommend that researchers look for general principles that can be applied regardless of the environmental conditions. In others, we encourage military researchers to take environmental factors into consideration when appropriate.

The issues we address in this chapter involve all aspects of the evaluation framework introduced in Chapter 1. Our framework will guide us to appropriate evaluation methods for a given situation or specific aspect of performance management. Because recruiter performance management systems including both existing and proposed new programs are concerned with recruiter attitude, intentions, and behavior, multiple evaluation methods will have a role. Thus, in evaluating performance management systems, there will be situations in which experiments, or econometric techniques, or qualitative techniques are most appropriate.

APPROACHES TO EVALUATION OF RECRUITER PERFORMANCE MANAGEMENT

In many ways, military recruiting is similar to other kinds of sales activities in the civilian sector. While the military recruiting environment includes many features that distinguish it from the civilian sales environment, it is not so unusual that knowledge gained in the civilian environment should be ignored. There is a substantial body of research on performance management of sales forces. When those research findings are coupled with an examination of best practices used in high-performing sales organizations, military recruiting management is presented with numerous alternatives to current practices. In choosing among those alternatives, recruiting management can design and carry out studies using the techniques described in earlier chapters.

In the remainder of this chapter, we describe some approaches that could be considered in evaluating the effectiveness of recruiter performance management alternatives. We review areas in which such approaches may be productive, not to describe in detail how such evaluations should take place. Many of the topics discussed—selection, training and development, reward and recognition programs, and performance assessment—have extensive research literatures that deal with many of the problems the Services face, and the Services should seek out appropriate expertise in designing and executing evaluations of such programs.

Recruiter Selection

The problem of selecting people who will become successful salespeople is not unique to the military, yet there are elements of this selection problem that are found only in the Services. One aspect of staffing the military recruiting function that presents a substantial challenge is the pipeline of potential recruiters from which the military chooses. The Services have specific needs for recruiting personnel and have generally concluded that uniformed Service personnel will be detailed into recruiting positions. The pool of potential recruiters thus consists of both volunteers and nonvolunteers, who have already been trained in some other military specialty and have some track record of success in that other specialty. In some respects, then, the military may know more about its potential recruiters than most organizations know about their candidates for sales positions. However, organizations hiring salespeople can look for people with previous (successful) sales experience and can assume that the vast majority of applicants actually desire to be hired. Neither the strategy of looking for previous experience in sales nor the assumption of motivated applicants is available to the military.

Although the Services have traditionally used enlisted personnel as recruiters, Congress recently mandated experimental use of civilian recruiters in the National Defense Authorization Act for Fiscal Year 2001. These experiments are not yet complete; however, the results of such studies should be taken into account when selection issues are considered.

We have previously noted that the Services today do not always give major weight in the recruiter selection process to a candidate's potential for success in a sales environment (National Research Council, 2003). Similarly, the U.S. General Accounting Office noted that the Services typically focus on past job performance in nonrecruiting (i.e., nonsales) positions when selecting recruiters (U.S. General Accounting Office, 1998). Thus, more efficient and effective methods for choosing specific personnel who should be assigned as recruiters are likely to exist.

There is a substantial literature, both military and civilian, addressing the problem of selecting people for sales occupations generally and military recruiting positions specifically. For example, Vinchur, Schippmann, Switzer, and Roth (1998) reviewed 97 studies of the relationship between predictors and job performance of salespeople. They conclude that personality dimensions, tests designed specifically for predicting sales success, individual interests, and other biographical information items are useful in selecting people for sales occupations. Recent reviews of the literature discuss many possible approaches that could increase the likelihood of successfully selecting recruiters who will be high producers (Penney, Horgen, and Borman, 2000; Penney, Sutton, and Borman, 2000b; McCloy et al., 2001). Similarly, there is substantial guidance in the professional literature on appropriate ways to evaluate the effectiveness of selection systems (American Psychological Association, 1987).

The process for determining the effectiveness of a selection procedure is relatively straightforward and well understood. Although there are several approaches to establishing the effectiveness selection procedures (i.e., the extent to which a selection procedure predicts all or part of required job performance), a common approach involves statistically relating test scores and job performance measures. Researchers begin by conducting a job analysis—often employing multiple methods, such as focus groups, interviews, and surveys—and using its results to identify the tasks that are performed by incumbents and the knowledge, skills, abilities, and other characteristics (KSAOs) necessary to perform those tasks. They then search for existing measures (such as tests and interviews), adapt existing measures, or develop new measures of those KSAOs. Previous research indicates that cognitive ability, various personality traits, and vocational interests may be appropriate constructs to be measured in recruiter selection tools. Often, a combination of such measures proves to be the best predictor of job performance (Borman, Toquam, and Rosse, 1978). Candidate selection measures are administered to some number of applicants for a position (or incumbents on a job), performance data are collected after some period of time on the job, and the relationship between performance on the measures and performance on the job is determined. Experimental (or quasi-experimental) designs are used to ensure that inferences from the study can be interpreted.

Managers should be aware that they need not invest immediately in multiple, large-scale studies that disrupt their routine processes in order to sort through the many options that are available for selecting recruiters. Small-scale studies can be very helpful in eliminating options or identifying those options with high promise. In addition, some of the Services have pursued this line of research extensively.

McCloy et al. (2001) provide a recent example that implements this general approach. They focused on estimating the value of a cognitive abilities test (in this case, the Armed Services Vocational Aptitude Battery or ASVAB) and recruiter school grades as predictors of the quantity and quality of recruiter productivity. Their analyses controlled for additional factors that can affect recruiting productivity, such as the number of young men and women in the local population and the number of high schools in the vicinity. They then added an evaluation of the cost-effectiveness of a particular selection approach; that is, they considered the trade-off between the cost of implementing a selection system and the gains experienced from increased productivity if the system is implemented. McCloy et al. concluded that adding ASVAB scores and other available demographic information is a cost-effective method for improving the selection of Navy recruiters.

Two issues mentioned above must be addressed when carrying out this type of selection research. First, appropriate measures of recruiter performance are required. Choices available here illustrate the difficulty of determining which performance measure—or measures—to use. For example, number of contracts in some period of time might be appropriate and is often used as a starting point in the process. However, markets vary in size and propensity; thus, number of contracts is often adjusted to consider those factors. Similarly, not all contracts are alike. Some recruits are more desirable than others; thus, number of contracts is often qualified by specifying "high-quality" contracts. Finally, number of contracts clearly does not capture all aspects of a recruiter's job.

Once a contract is signed, administrative requirements must be attended to, and recruiters must continue to motivate applicants until they actually arrive at their basic training location. For example, Borman, Hough, and Dunnette (1976) identified eight dimensions of military recruiter performance: (1) locating and contacting qualified prospects; (2) gaining and maintaining rapport; (3) obtaining information from prospects; (4) salesmanship skills; (5) establishing and maintaining good relationships in the community; (6) providing knowledgeable and accurate information; (7) administrative skills; and (8) supporting other recruiters and the Recruiting Command. From this list, perhaps only dimensions 1 through 4 are being considered directly in a performance measure that is based on number of contracts. If dimensions 5 through 8 are also important dimensions of recruiter performance, then the number of contracts must be supplemented with other performance measures. In sum, appropriate performance measures must be carefully chosen to represent those performance dimensions being targeted by the proposed selection measure.

The second issue is relatively rare in the civilian selection arena, in which most applicants desire the job to which they are applying. Specifically, those people being considered for the job of military recruiter do not always actually want the job. Thus, the evaluation strategy described above will be useful for identifying measures that will help select volunteers who will be most productive, but those same measures may not be as useful in selecting among nonvolunteers.

People with the KSAOs to perform the job may not have the motivation to do so in an environment in which the job is particularly difficult and the rewards are low. Thus, the development and validation of selection procedures that identify individuals with the KSAOs to perform the job (particularly those procedures that focus on cognitive skills) may be inadequate if they do not identify those with the willingness to perform at the high levels needed by the Services in a perceived low-reward environment. Thus, in addition to considering the total context of recruiter performance, selection procedures must also better address the motivational aspects of job performance. It is important to remember that the selection problem should not be considered out of context. Military recruiting is not just another sales job. Moreover, the environment in which recruiters work can have a profound effect on motivation. To some extent, motivation to perform the job might be better addressed by manipulating the reward and recognition structure as well as the work environment in which the recruiter performs rather than by developing better selection tools.

Recruiter Training and Development

Once recruiters are selected, they must be trained to perform their jobs, and they must continue to be developed as they progress in their recruiting career. As with most training requirements, there are multiple strategies available to address the recruiter training requirement, and there are constrained training budgets that limit the amount of time (and thus the specific tasks) for training. As structured today, the Services offer a combination of full-time, in-residence training along with on-the-job training to try to ensure that recruiters have the skills required to be successful on the job. In evaluating the overall effectiveness of training, recruiting management must address two central issues. The first concern is the extent to which the training provided adequately covers the entire recruiting job. The second concerns the effectiveness of the training given. While not an area for research, a third issue related to effectiveness of training that should be mentioned is accessibility of training—ensuring that the right training is available at the right times in a recruiter's assignment.

Determination of training requirements typically begins with a needs analysis that specifies what skills are required to adequately perform a job and includes an analysis of where and how those skills might be acquired. Processes for both steps are well documented in the professional training literature. Unless the training is directed at developing the appropriate skills for the job, it is unlikely to be effective.

As with recruiter selection, there is a substantial professional literature addressing the issues involved in the development of recruiter training programs, both in the civilian environment and in the military. The committee has recommended that the Services develop and implement training systems that make maximum use of realistic practice and feedback (National Research Council, 2003). The Federal Advisory Committee on Gender-Integrated Training and Related Issues (1997) also included recommendations for improving recruiter training.

The second issue is evaluating training effectiveness. Is a specific training approach accomplishing its goals? How effective is it compared with alternative approaches? There is also an extensive professional literature regarding training evaluation in both the civilian environment and the military. While Salas, Milham, and Bowers (2003) note that rigorous training evaluations are rarely completed for military training courses, that does not mean that such evaluations should not be completed or cannot be completed. On the contrary, training evaluation is usually cited as a critically important aspect of the instructional systems design process used by the Services.

The need to evaluate recruiter training programs has not been ignored. Hull and Benedict (1988) proposed an evaluation methodology for Army recruiter training; Hull, Kleinman, Allen, and Benedict (1988) carried out an evaluation based on that methodology. Those authors used outcome variables that included ratings from instructors and both current and former students. Chonko, Madden, Tanner, and Davis (1991) used a qualitative approach to evaluate the effectiveness of the specific sales techniques taught to Army recruiters. Kirkpatrick (1976) described four levels of training evaluation: (1) reaction, (2) learning, (3) behavior change, and (4) results.

When Salas et al. note the lack of rigorous training evaluations, they are focusing on evaluations at levels 3 and 4 of the Kirkpatrick model. Frequently, survey methods are used to determine whether trainees like (or dislike) a particular training course, and end-of-course tests (coupled with tests given at the beginning of a course) determine whether trainees increased their knowledge in a subject area. While trainee reactions and mastery of specific knowledge are both important aspects of military recruiter training (or sales training in general), more critical outcomes of this training include whether or not the trainee can apply what he or she has learned (behavior change) and whether or not the trainee becomes an effective recruiter (results).

Appropriate evaluation of training programs often requires the use of multiple strategies and techniques. As noted above, survey techniques may be used as a component of training evaluation to determine trainee and instructor reactions to the training. Surveys can determine whether trainees liked the training, believed that the training improved their skills, believed that the training would be applicable on their jobs, believed that time devoted to particular tasks was appropriate, and so forth. Similarly, surveys of trainees and their supervisors provide information useful in guiding training course design and revision. Surveys can be used to determine whether supervisors believe that the training syllabus includes the proper tasks or skills that trainees need, that time devoted to particular tasks is appropriate, and so forth.

At the same time, experimental and quasi-experimental designs play roles in evaluating whether training results in a specific behavior change or a change in the overall effectiveness of job performance. While it may be inconvenient or expensive to implement true experimental designs (with treatment and control groups and random assignment to conditions), preexperimental or quasi-experimental designs can provide valuable information in the overall evaluation of training programs (Sackett and Mullen, 1993). Outcome measures for these studies must be carefully chosen. We noted the Borman et al. (1976) dimensions of recruiter performance in our discussion of recruiter selection issues. The choice of appropriate criteria against which to evaluate recruiter training is a similarly complicated issue. If the eight Borman et al. dimensions are all determinants of successful recruiter performance, then it follows that they are all candidates for inclusion in the evaluation of recruiter training programs. Choosing only a single outcome measure, such as number of new contracts in some fixed period of time, as the standard for evaluating recruiter training that covers all aspects of the recruiter job provides little information that would be useful in improving the training course.

It is important to remember that development takes place in ways other than formal training programs. Often, individual feedback and coaching around certain experiences is a very effective way to shape behavior. Experiential learning and associated coaching assume that there are capable coaches who understand what the desired behavior is and who can communicate performance deficits and strategies for improvement. Thus, several potentially productive activities may include the Services studying the informal development of recruiters and identifying ways to ensure that those who supervise recruiters possess adequate knowledge of the recruiting process as well as coaching skills themselves.

Recruiting Office Locations

Once the Services have selected and trained recruiters, the question arises, "Where should recruiters be located?" As with training, the answer to this question is always constrained—by budget, by policies that dictate combining offices from multiple Services when feasible, etc. Along with the location of each specific office, there are also questions addressing the number of recruiters who should be stationed in that office. Each of these decisions is open to evaluation, with the goal of establishing the most effective recruiting organization for a given budget, or with the goal of minimizing the cost of obtaining a given number of recruits.

There is little systematic research on strategies for locating recruiters. However, existing data lend themselves to econometric analyses. For example, recruiter productivity over time should be available for existing offices. That productivity can be modeled by such available data as size of qualified recruiting market in the office's geographic region, experience levels of the recruiters assigned to that office, incentive programs (for recruits or for recruiters) that were available at given times, and so forth. Such analyses could guide recommendations for appropriate placement of recruiting offices. Through econometric modeling, the Services should minimally be able to define the relevant variables for establishing recruiting locations and to consider various methods for determining optimal staffing levels.

Recruiter Production Goals

Most sales organizations establish goals (targets or quotas) for performance of their sales force. Many variables potentially shape these goals. For example, some organizations take into account the "product" being sold as well as the area in which it is sold and past demand for the product. Some may even take into account the experience and expertise of the salesperson. Military recruiting is no exception to this model. The question of how best to establish recruiter goals (and the question of whether those goals should be individual or team based) is still open. Given that recruiting duty is often cited as an extremely stressful job—due to the constant pressure to "make goal"—there would seem to be high payoff in defining the variables relevant to military recruiting goals and evaluating the goal-setting process.

Recruiter Reward and Recognition Programs

Sales organizations generally have well-established reward and recognition programs, closely tied to performance as measured against goals. Here as well, military recruiting organizations are little different from their civilian sales organization counterparts (although military recruiting organizations must contend with legal constraints against using financial incentives for recruiters). The hallmark of many sales reward and recognition programs is pay and bonuses based on sales performance. Pay for recruiters is determined by Congress and cannot be adjusted for this job. Because the military cannot use most types of financial incentives, the Services rely instead on nonfinancial incentives such as plaques, watches, rings, and military decorations in lieu of cash compensation. However, the effectiveness of such incentives as a reward for past performance and a subsequent motivator of future performance remains an open question. The Federal Advisory Committee on Gender-Integrated Training and Related Issues (1997) made a number of recommendations for changes in recruiter incentive systems. For example, they suggest that the overall level of recruiter incentives needs to be increased, that the Services should make a recruiting assignment career-enhancing, and that a recruiter's rewards and recognition should be linked to his or her recruits' performance in basic training.

All of the evaluation approaches raised in this report should be considered when attempting an evaluation of these incentives. For example, there will often be adequate existing data to support econometric analyses of the effects of specific programs on recruiter behavior as measured by productivity over some period of time. Asch (1990) provides such an example. She used demographic data and varying incentive structures to estimate effects on effort and productivity, concluding that rewards and the timing of rewards affect the allocation of recruiter effort over time. In the absence of existing data, experiments could be devised to determine the effect of alternative programs on recruiter intentions (as measured by surveys) or recruiter behavior (as measured by productivity).

In addition to evaluating the effectiveness of existing incentives, the Services must also consider other approaches to reward and recognition. Defining the most appropriate array of alternatives must be based on review of the existing compensation literature and analysis of the needs and expectation of the recruiting work force. The Services could use focus groups and interview techniques to generate a list of alternative incentives that might be expected to motivate recruiter effort and understand the trade-offs among them. For example, the Services may find that recruiters are willing to accept low cash incentives if sales (i.e., recruiting performance) are rewarded by career enhancement.

The Services could also explore the acceptability and potential incentive value of various career enhancements. We note the widespread perception that successful duty as a recruiter or drill instructor is necessary for advancement to senior noncommissioned officer rank in the Marine Corps. In addition, there have been suggestions that recruiter special duty assignment pay—which is now based exclusively on length of time as a recruiter—could be structured to provide an incentive for production rather than an incentive for staying in the recruiting job.

Interestingly, there are some financial disincentives associated with the recruiter's job. Anecdotally, recruiters often recount the financial hardships of living in areas without a substantial military presence. The lack of services often provided on military bases and the cost of medical care are often cited. Thus another approach to rewarding recruiters is removing the financial disincentives.

Recruiter Performance Assessment

Answers to any of the questions raised above—from recruiter selection through recruiter reward and recognition programs—assume that there are some measures of recruiter (or team or organizational) performance against which alternatives can be compared. These performance assessment measures themselves are also open to evaluation in a number of ways.

First, the Services must define the critical aspects of the recruiter job. The Borman et al. (1976) study defined eight dimensions of recruiter performance; however, we must note that the study is almost 30 years old. Since many things about the Services and the recruiting environment have changed in the past 30 years, we could reasonably assume that the recruiting task has also changed. In addition, enlisted military recruiters are also part of their respective Services. Those dimensions of performance associated with simply serving must also be included if the performance assessment is intended to cover the entire job.

A second task is to assess the accuracy of the appraisal itself. Numerous studies have noted that military performance measures tend toward leniency—that is, disproportionately large numbers of individuals receive high ratings. Similar problems exist in civilian performance appraisal systems. The use of experimental performance measures in test development and validation work also suggests that typical measures are not adequate. Consequently, the Services must continue to develop and employ performance appraisal systems that provide accurate ratings of job performance.

The military has already devoted considerable resources to the development of performance appraisal systems. Theoretically, at least, the problem may not be with the performance appraisal system but with the environment in which it is used. Thus, a third avenue of exploration is studying the conditions under which performance assessments are made and used. Because the recruiting job is so unlike other military jobs, the Services might take the approach of developing a performance appraisal system that is outside the routine systems used for administrative purposes. Such a system might have feedback to the recruiter as its primary purpose, rather than rating or ranking recruiters.

Another area of research the Services may find productive is to link performance assessment to training and development options. If the Services are able to identify specific areas of weaknesses and provide a feedback mechanism, then follow-up suggestions for remediation will be required to actually improve the skills of the recruiting workforce. Experimental research will be necessary to identify development opportunities that actually rectify performance deficits.

A final area of evaluation is the effectiveness of communications about the performance appraisal system. Stating what is to be rewarded is one step in increasing the probability that such behavior will be exhibited. If performance expectations are greater than simply the number of contracts, then that information needs to be clearly communicated to recruiters so that they can manage their performance in light of a broader view of performance. Similarly, comprehensible, individual feedback on performance is another critical step in ensuring that recruiters understand their own strengths and weaknesses and are able to target their own developmental efforts appropriately.

CONCLUSIONS

In many respects, the problems of performance management faced by the military are no different from the problems faced by private industry. However, the environments are distinctly different, and the military faces many restrictions to which the private sector is not subject. Because of these environmental differences, some of the existing research from the professional literature will be useful, some not. Ideally, the military should undertake continuous and systematic evaluation of each aspect of performance management individually and as a whole in order to improve recruiter performance, relying on professional literature when possible, undertaking its own research when necessary.

Research that considers the interactions among various factors that affect recruiting performance is particularly needed. However, to undertake such massive research may be both overwhelming and practically and economically infeasible. Nevertheless, doing nothing heightens the likelihood of recruiting problems ranging from difficulty assessing the more qualified candidates to expenditures disproportionate to the results achieved. The challenge for the Services will be to establish their own overall research frameworks, prioritize their many options, select those research options that are most promising, and continually revise the research plan based on findings and changes in the environment.

9

Conclusions and Recommendations

The primary objective of this study is to help the Department of Defense (DoD) improve its research on advertising and recruiting polices. We anticipate that in the coming decade DoD will field and test new advertising and recruiting initiatives designed to improve the recruiting outlook and avoid the shortfalls of the last decade. In order to discover the most promising policies, in the committee's view DoD needs a comprehensive research and evaluation strategy based on sound research principles that will ensure valid, reliable, and relevant results. In this report, we present an evaluation framework that links different types of research questions to different research methodologies.

The framework identifies four major categories of research questions and four broad methodological approaches. The first category of research question asks: "What does a target audience see as attractive or unattractive features of a program?" It is well suited to examination via qualitative methods, such as focus groups; unstructured or open-ended surveys; and interviews. The second category of research question asks: "What is the effect of a program on specified attitudes or behavioral intentions?" It is well suited to examination via surveys, experiments, and quasi-experiments. The third category of research question asks "What is the effect of a proposed new program on enlistment?" It is well suited to examination via experiments and quasi-experiments. The final category of research question asks "What is the effect of an existing program on enlistment?" It is well suited to examination via econometric modeling.

The committee's work during Phase I led us to conclude that there are a number of critical problem areas or topics needing more intensive study.

Some problems arise because of the need for ongoing, up-to-date information that can serve as early-warning indications of potential recruiting problems or that can point to areas in which improvements are needed. Other problem areas are important because, in our view, they are central to improving the overall recruiting climate. We selected six areas as the central focus of this report. After devoting Chapter 2 to issues of theory as a guide to effective evaluation research, Chapters 3 through 8 each examine one of the six areas. The various chapters also introduce different methodological approaches to evaluation.

CHAPTER 2: THEORETICAL APPROACHES

The chapter outlines a general framework for thinking about effective program design. The first step is to identify the fundamental factors that impact a target population's enlistment behavior. The second step is to derive strategies (often informationally based) to change, enhance the effect of, or mitigate the effect of those determinants. We outlined a wide range of variables and processes that program designers must potentially take into account, drawing heavily on research from adolescent development, communications, economics, psychology, and sociology. These perspectives set the stage for conducting the necessary research to inform program design and program evaluation.

There are two distinct theoretical approaches to enlistment behavior: decision theory, based primarily in psychology, and the econometric theory of enlistment supply. While there is some overlap in these two theoretical traditions, they have distinct approaches. Decision theory is more highly developed for the purpose of conceptualizing and measuring behaviors that affect individual decision making. Econometric theory is more formally developed with respect to aggregate enlistment outcomes and various exogenous influences. Accordingly, a key objective of this chapter is to build an integrated perspective on the behavioral and econometric approaches.

Conclusion: The role of theory is crucial to the design of interventions to increase enlistment behavior. When enlistment programs are developed atheoretically, they run a great risk of being ineffective.

CHAPTER 3: MONITORING TRENDS IN YOUTH ATTITUDES, VALUES, AND PROPENSITY

Determinants of Propensity

The chapter proposes survey methods as the most suitable research design for tracking changes in propensity as well as for assessing the underlying beliefs that are related to propensity. The information gathered in youth attitude surveys is valuable to many research studies, such as the advertising studies proposed in Chapter 4, but it is also valuable in its own right by providing early warning indicators of changes in the propensity for military service. The chapter reiterates a key point from the committee's earlier report, namely, that as propensity to enlist is the major direct determinant of actual enlistment, increasing propensity in the youth population should be a key objective for the military. We summarize a model of the determinants of propensity that we laid out in the earlier report and build an argument that research on propensity in the youth population should measure the key determinants of propensity. We briefly review ongoing survey efforts dealing with propensity, noting that they do not consistently include these key determinants of propensity. We also note a tendency for survey research dealing with propensity to make use of research designs that do not yield complete data on individuals. If the interest is simply documenting the proportion of respondents choosing each alternative to a survey item, then randomly distributing items among respondents will yield accurate results, as individual-level analysis is not central to the research question. But when there is interest in the pattern of relationships among variables, designs involving complete data at the individual level are needed.

Conclusion: Previous survey research examining propensity to enlist has not consistently measured the key classes of determinants of propensity (i.e., attitudes, norms, and self-efficacy), nor has it consistently used research designs permitting analysis at the individual level. This has limited the ability to test complete models of the determinants of propensity.

> **Recommendation:** We recommend that survey research examining propensity be designed to incorporate the key determinants of propensity and that it be designed to permit meaningful analysis at the individual level.

Conclusion: We also note that our model of the determinants of propensity includes the role of important influencers, such as parents and peers. One important implication of this is that the effects of interventions, such as advertising, are not fully addressed by focusing solely on the direct effects of the intervention on the potential recruit.

> **Recommendation:** We recommend that evaluation efforts consider potential effects on key influencers as well as on potential recruits, and that efforts be made to assess such indirect effects on propensity.

A Program of Survey Research

The second part of the chapter provides a series of concrete recommendations for a program of survey research, expanding on ideas set forth in our letter report to the Department of Defense of June 2000. A program of monitoring surveys, which have the potential to yield very high-quality data about propensity and its determinants, is presented. We propose a cohort-sequential design, in which samples of youth are obtained annually in the 11th grade of school (i.e., age 16-17) and monitored through the age of roughly 23. We discuss a range of issues in the design of such a project, including information-gathering format (e.g., self-completed questionnaire versus telephone interview), means of accessing the sample (e.g., school-based surveys versus random household sampling), mechanisms for follow-up surveys over time, issues in the scheduling of surveys, and sampling strategies. We note that such a project is a significant investment and should not be undertaken unless the resources for a minimum of five years can be committed. We note that a variety of options are available with the broad framework we develop. For example, one possibility is to survey 11th graders, with annual follow-up; another is to survey both 11th and 12th graders, with follow-up every two years. The chapter develops the trade-off among the options, noting that many details cannot be specified in advance.

> **Recommendation:** We recommend that consideration be given to undertaking a school-based survey, using a cohort sequential design, in which students are sampled in the 11th grade and possibly the 12th grade and regularly resurveyed until the age of 23 or 24.

Item Content

The final section of the chapter discusses item formats for assessing the important variables in the model of the determinants of propensity developed in our earlier report and reiterated here. Specific examples of item formats for effectively measuring propensity, attitude, norms, self-efficacy, behavioral beliefs, outcomes, and open-ended queries about outcomes and about influencers are provided.

Recommendation: We recommend that surveys dealing with propensity and its determinants assess the variables of interest using established item formats.

CHAPTER 4: ADVERTISING PLANNING: GENERATIVE AND EVALUATIVE APPROACHES

In order to develop and test effective advertising themes, two types of research designs are required. The first type of design is for the development of preliminary but promising message strategies. This step requires qualitative exploratory or generative research designs (focus groups, in-depth interviewing, etc.). After promising themes are developed, the second type of research design is necessary for testing theme awareness and market impact. The best designs for this step are experimental and quasi-experimental studies.

This chapter describes the stages in the development and evaluation of an advertising campaign, discussing relevant research methods for each phase. The chapter reviews some findings from the committee's earlier report, including the decline in both the proportion of youth assigning high value to duty to country and in the proportion of youth who associate the goal of duty to country with military service. We then develop a framework for developing advertising campaigns that follows a systematic process and builds on sound information about the value structure of youth. That framework involves (1) tracking the competitive environment for military recruitment to detect factors affecting youth understanding and views of military service; (2) examination of audience member beliefs, goals, and language; (3) development of a range of message strategies for military recruitment, and (4) allocation of resources to advertising message strategies.

Conclusion: There is a need for research to provide a more complete picture of the belief and value structure of the youth population, particularly the beliefs relating to public service, duty to country, personal sacrifice, and concern for others. It is also helpful to study the language used by youth as they think and speak about these issues, so that this information can be used to develop effective messages on the topic.

Recommendation: We recommend a program of research that begins with generative techniques to understand the concepts and language used by youth in considering alternative courses of action (e.g., education versus military service) and continues with survey research that measures the full range of beliefs, attitudes, and values that emerge as linked to these alternate courses of action.

Conclusion: Effective advertising campaigns involve a message strategy strongly linked to beliefs and values that affect decision making. A crucial component of the evaluation of military advertising is an examination of its success in affecting the intended values and beliefs. Using beliefs and values as outcome variables as well as enlistments permits a clearer understanding of why a given advertising campaign is or is not successful than using enlistments alone.

Recommendation: We recommend that advertising message strategies be evaluated in terms of their effects on targeted beliefs and values. Such evaluation should make use of experimental designs in controlled settings as well as small-scale, in-market experiments.

Recommendation: We recommend that a policy be adopted of regularly developing and evaluating alternative approaches that challenge existing message strategies.

CHAPTER 5: DETERMINING OPTIMAL LEVELS OF ADVERTISING AND RECRUITING RESOURCES

The chapter focuses on econometric methods, as these approaches are most useful for assessing the optimal levels of recruiting programs and resources. Econometric methods can be used to isolate and identify the effects of existing resources, policies, and external factors affecting recruiting outcomes as well as their costs. There is by now a relatively well-developed body of econometric research that has identified some of the most important determinants of enlistment supply as well as the cost and effectiveness of various trade-offs among different recruiting and advertising resources. Estimates are based on the natural variation in key recruiting resources and outcomes (usually aggregated) that occur over time and across different geographic locations.

Brief overviews of two types of econometric models are provided—models of enlistment supply and models of recruiting cost. We review studies over the past two decades on the effects of recruiting and advertising on enlistment and present summary tables comparing the various studies. We note considerable variability in results across studies and suggest a series of methodological features that have not been consistently incorporated into the studies and thus may contribute to the variability in results and the difficulty in giving a definitive answer to questions about the elasticity of enlistment with respect to advertising.

Conclusion: More sophisticated methods, controlling appropriately for factors affecting enlistment supply, both those that are directly observable to the researcher and those that must be inferred, such as recruiter effort, are necessary to obtain efficient, unbiased estimates of the effects of recruiting resources. Moreover, more complete evaluation of the effects of some types of resources, especially advertising content, require estimation using more flexible functional forms in the econometric analysis. To apply these methods, however, better data need to be collected and systematically maintained. The specific conclusions and recommendations discussed below are conditional upon the availability of better data.

> **Recommendation:** Collect and maintain better data to support the estimation of enlistment supply functions and to evaluate the effectiveness of recruiting resources.

Conclusion: Recruiter productivity varies with experience, and hence sudden changes in the size of the recruiting force result in declines in average experience. Failure to incorporate recruiter experience in models of recruiter effects may bias study results.

> **Recommendation:** We recommend that future research on the effects of recruiters on enlistment supply incorporate the effects of recruiter experience.

Conclusion: Recruiter incentives have been incorporated in supply models via recruiters' quotas, based on the assumption that increasing recruiting quotas increases effort.

> **Recommendation:** We recommend that supply models incorporate more complete and realistic models of recruiter incentives that more fully capture the complexities of recruiter incentives.

Conclusion: Research to date has not incorporated the effects of reserve forces competition on active-duty recruiting.

> **Recommendation:** We recommend that supply models incorporate reserve forces competition for nonprior-service recruits.

Conclusion: Econometric estimates of the effects of advertising have focused largely on expenditures of impressions (i.e., the number of relevant individuals who see or hear the advertisement, often measured in terms of gross rating points). Such estimates have not attempted to measure differences in effects by specific advertising content.

> **Recommendation:** We recommend that research attempt to evaluate advertising in terms of thematic content in order to determine whether advertising effects vary by content, as well as by impressions and expenditures.

Conclusion: The functional forms (i.e., the shape of the relationship between the recruiting incentive and enlistment) of econometric supply models have been relatively restricted. The underlying assumptions (e.g., that each additional advertising dollar has the same effect regardless of the level of total expenditure) may not be correct, and an examination of more flexible functional forms would be fruitful.

> **Recommendation:** We recommend that research on supply models make use of flexible functional forms, rather than imposed functional forms.

CHAPTER 6: THE TIMING AND LEVELS OF JOINT AND SERVICE-SPECIFIC ADVERTISING

Minimum Advertising Level to Maintain Propensity

The chapter first addressed the question of whether there is a minimum level of advertising necessary for a cost-effective recruiting program, even if that advertising is not necessary to achieve contemporaneous enlistment contract goals. Historically, when the recruiting climate is good and recruits are plentiful, military planners tend to cut advertising budgets, thereby contributing to a reduction in awareness capital and propensity levels. This may possibly set up a boom or bust cycle, in which propensity falls, recruiting becomes more difficult, and then advertising funds have to be restored. We present a model that describes the conditions under which it would be cost-effective to advertise in the interests of future enlistment supply and review research to date that speaks to the issue. While extant research suggests that advertising may have effects for only a short period of time, the data available to prior researchers are limited for several reasons. First, they do not permit examining both lagged effects and nonlinear effects within a time period. Second, they focus on advertising aimed at youth at the point of the enlistment decision and do not permit examining possible supplemental advertising approaches, such as those aimed a youth several years prior to an enlistment decision, or those aimed at adult influencers, such as parents.

Conclusion: Research to date does not permit a definitive answer to the question of the cost-effectiveness of advertising above and beyond that which is necessary to achieve current recruiting goals.

> **Recommendation:** We recommend a focused effort to maintain advertising data in a systematic way for purposes of estimating a supply curve that incorporates the potential for both time-lagged and nonlinear advertising effects.

Recommendation: We recommend a program of research, incorporating quasi-experimental methods, to examine advertising effects over an extended period of time.

Levels of Joint and Service-Specific Advertising

The chapter then turns to the optimal levels of joint and Service-specific advertising. It is our opinion that certain types of advertising themes, such as generic themes designed to increase overall propensity, are best done as a joint program, while advertising themes featuring specific benefits of military service are best done in the Service program. What we do not know is what advertising fund level should be allocated to joint programs. We note that issues of scale play a role in addressing this issue, as certain types of advertising (e.g., television) do not appear to have a constant effect across levels of expenditure. The larger the recruiting effort and the larger the budget, the greater the potential value of a multifaceted campaign, with some resources targeted toward providing information about specific Services to those already with a propensity to enlist and others targeted toward increasing propensity among those currently without it.

Recommendation: We recommend a program of research aimed at examining the effects and cost-effectiveness of information-oriented versus values-oriented advertising in joint and Service-specific advertising programs.

CHAPTER 7: DETERMINING OPTIMAL TYPES OF INCENTIVES

Over the years of the All-Volunteer Force, various incentives have been developed and offered to help strengthen and shape military enlistments. The effectiveness of these incentives has been addressed, and demonstrated, using a variety of evaluation approaches. This chapter considers methodological issues in determining the optimal types of enlistment incentives for specific recruiting goals. We begin by reviewing different types of enlistment options. Next, we consider the various types of effects the incentives may produce, both intended and unintended, and the related methodological issues in assessing their impact. We continue with a brief review of the analysis methods discussed in the earlier chapters, which is followed by a longer consideration of analytical issues applicable to each of the evaluation approaches. Finally, drawing on each of these areas, we conclude with a discussion of matching potential incentives and their effects with the appropriate assessment goals and evaluation methods.

A central message of the chapter is that each of the evaluation methodologies introduced in this volume (qualitative methods, surveys, econometric models, and experiments and quasi-experiments) can play a useful role in addressing different questions that policy makers may ask about current or proposed incentives. The chapter offers concrete illustrations of the research framework introduced in Chapter 1. It identifies focus groups as of particular value in providing insight into the appeal of various features of proposed incentives and in facilitating the discovery and exploration of new incentive options. It identifies surveys as of particular value in comparing the relative attractiveness of a substantial number of incentive options. It identifies econometric methods as of particular value when examining the effects of existing programs on actual enlistments over time and over differing recruiting environments, providing statistical control for a wide variety of potentially confounding factors (e.g., geographic effects, effects of changes in the economy). It identifies experimental and quasi-experimental methods as of particular value when the question of interest is estimating the effects on enlistment of a new incentive prior to full implementation.

In addition, the chapter emphasizes the value of combining approaches. This might include, for example, the use of focus groups to help explore and define prospective enlistment options that are then tested in large youth market surveys or in pilot tests employing experimental designs. Another example would be the application of econometric methods or other analytical techniques to existing datasets in order to help interpret survey results; for example, the other approaches could be used to derive a metric that can be used to help translate stated enlistment intentions in the survey into estimated enlistment rates.

The focus of the chapter is not on specific conclusions and recommendations, but rather on illustrating the range of available options for evaluating incentives, making a case for the linkage of research methods chosen to the research question of interest and advocating for a combination of research methods as appropriate.

CHAPTER 8: PERFORMANCE MANAGEMENT OF RECRUITERS

This chapter shifts the focus from influence attempts aimed at the potential recruit to examination of recruiting systems. Service recruiting managers establish systems to select recruiters from among the available pool of Service members, to train and develop those new recruiters, to open recruiting offices in specific locations, to establish production goals for each recruiter, to motivate recruiters with reward and recognition programs, and to monitor and assess recruiter performance. Many options are available for each of these systems, and each is open to evaluation. In some cases (for example, selection of new recruiters), there are continuing research programs to evaluate the effect of alternative programs. In other cases (for example, the effects of recruiter reward and recognition programs), research or evaluation is rarely attempted. Given the central role that recruiter productivity plays in the recruiting process, all aspects of recruiter performance management should be subjected to evaluation efforts.

Recruiter Selection

There is a substantial literature, both military and civilian, addressing the problem of selecting people for sales occupations generally and military recruiting positions specifically. There is a long history in civilian settings of successfully utilizing various selection techniques to identify individuals with a high likelihood of success in sales-oriented positions. Selection in civilian settings involves an applicant pool eager to be selected, which is often not the case in the recruiter selection setting. It would appear worthwhile to consider changes and enhancements in the incentives to take on a recruiter position (e.g., links to career advancement) in order to increase the pool of individuals able and willing to serve as recruiters.

Conclusion: Given the body of research on selection for sales-oriented positions, it is likely that there are more efficient and effective methods than currently used for choosing those personnel who should be assigned as recruiters.

> **Recommendation:** We recommend continued research on the development of effective recruiter selection strategies, in conjunction with a consideration of career incentives for service as a recruiter.

Recruiter Training

The committee's earlier report recommended that the Services develop and implement training systems that make maximum use of realistic practice and feedback. We note here the importance of evaluation of training programs, including giving careful attention to the outcome variables of interest.

It is important to remember that development takes place in ways other than formal training programs. Often, individual feedback and coaching around certain experiences are very effective ways to shape behavior. Experiential learning and associated coaching, however, assume that there are capable coaches who understand what the desired behavior is and who can communicate performance deficits and strategies for improvement.

Recommendation: We recommend that the Services expand their evaluation of overall training of recruiters to include the study of other informal development opportunities. In particular, assessment and improvement of the supervisory and coaching skills (to include on-the-job training) of those who train recruiters may be a fruitful approach.

Performance Goals

The question of how best to establish recruiter goals (and the question of whether those goals should be individual or team based) is still open. Given that recruiting duty is often cited as an extremely stressful job—because of the constant pressure to "make goal"—there would seem to be high payoff in defining the variables relevant to military recruiting goals and evaluating the goal-setting process.

Recommendation: We recommend a program of research aimed at evaluating the effects of goals on recruiter behavior and outcomes.

Recruiter Performance

Simple outcome measures (e.g., number of contracts) may be subject to a variety of external constraints (e.g., location) and may not capture the full range of important recruiter activities. There is a foundation of previous research on the dimensions of effective recruiter performance that merits updating. A complete and current model of the dimensions of recruiter performance is needed as the basis for an effective performance evaluation system.

Recommendation: We recommend research to develop a complete model of recruiter performance and to develop performance appraisal instruments and feedback processes based on this model.

References

Ajzen, I. (1985). From intentions to actions: A theory of planned behavior. In J. Kuhl and J. Bechmann (Eds.), *Action control: From cognition to behavior* (pp. 11-39). New York: Springer-Verlag.

Ajzen, I. (1991). The theory of planned behavior. *Organizational Behavior and Human Decision Processes, 50,* 179-211.

Ajzen, I., and Fishbein, M. (1980). *Understanding attitudes and predicting social behavior.* Englewood Cliffs, NJ: Prentice-Hall.

Albarracin, D., Johnson, B.T., Fishbein, M., and Muellerleile, P. (2001). Theories of reasoned action and planned behavior as models of condom use: A meta-analysis. *Psychological Bulletin, 127,* 142-161.

American Psychological Association. (1987). *Principles for the validation and use of personnel selection procedures.* College Park, MD: Society for Industrial and Organizational Psychology.

Anderson, N.H. (1996). *A functional theory of cognition.* Hillsdale, NJ: Erlbaum.

Asch, B.J. (1990). Do incentives matter? The case of Navy recruiters. *Industrial and Labor Relations Review, 43*(3), Special Issue, 89S-106S.

Asch, B.J., and Orvis,B. (1994). *Recent recruiting trends and their implications* (MR-549-A/OSD). Santa Monica, CA: RAND Corporation.

Asch, B.J., and Warner, J.T. (2001). Compensation and personnel management in hierarchical organizations: Theory and application to the U.S. military. *Journal of Labor Economics, 19*(3), 523-562.

Bachman, J.G., Freedman-Doan, P., Segal, D.R., and O'Malley, P.M. (2000). Distinctive military attitudes among U.S. enlistees, 1976-1997: Self-selection versus socialization. *Armed Forces and Society, 26*(4), 561-585.

Bachman, J.G., Segal, D.R., Freedman-Doan, P., and O'Malley, P.M. (1998). Does enlistment propensity predict accession? High school seniors' plans and subsequent behavior. *Armed Forces and Society, 25*(1), 59-80.

Bailey, R.M., Strackbein, M.E., Hoskins, J.A., George, B.J., Lancaster, A.R., and Marsh, S.M. (April 2002). *Youth attitudes toward the Military: Poll one* (DMDC Report 2002-027). Arlington, VA: Defense Manpower Data Center.

Bakken, T. (2002). The role of human agency in the creation of normative influences within individuals and groups. *Journal of Human Behavior in the Social Environment, 5,* 89-104.

Bandura, A. (1991). Self-efficacy mechanism in physiological activation and health-promoting behavior. In J. Madden (Ed.), *Neurobiology of learning, emotion, and affect* (pp. 229-269). New York: Raven.

Bandura, A. (1994). Social cognitive theory and exercise of control over HIV infection. In R.J. DiClemente and J.L. Peterson (Eds.), *Preventing AIDS: Theories and methods of behavioral interventions* (pp. 25-29). New York: Plenum Press.

Barnes, J.D., Dempsey, J.R., Knapp, D.J., Lerro, P.A., and Schroyer, C.J. (1991). *Summary of military manpower market research studies: A technical report* (FR-PRD-91-08). Alexandria, VA: Human Resources Research Organization.

Becker, M.H. (1974). The health belief model and personal health behavior. *Health Education Monographs, 2,* 324-508.

Becker, M.H. (1988). AIDS and behavior change. *Public Health Reviews, 16,* 1-11.

Berner. K., and Daula, T. (1993). Recruiting goals, regime shifts, and the supply of labor to the Army. *Defence Economics,* 4(4), 315-328.

Blanton, H., and Christie, C. (2003). Deviance: A theory of action and identity. *Review of General Psychology, 7,* 115-149.

Bogart, L. (1986). *Strategy in advertising: Matching media and messages to markets and motivation* (pp. 366-371). Lincolnwood, IL: NTC Business Books.

Bohn. D., and Schmitz, E. (1996). *The expansion of the Navy College Fund: An evaluation of the FY 1995 program impacts.* Arlington, VA: Navy Recruiting Command.

Borman, W.C., Hough, L.M., and Dunnette, M.D. (1976). *Development of behaviorally based rating scales for evaluating the performance of U.S. Navy recruiters* (NPRDC Technical Report 76-31). San Diego: Navy Personnel Research and Development Center.

Borman, W.C., Toquam, J.L., and Rosse, R.L. (1978). *Development and validation of an inventory battery to predict Navy and Marine Corps recruiter performance* (Institute Report 22). Minneapolis: Personnel Decisions Research Institutes.

Borsari, B., and Carey, K.B. (2003). Descriptive and injunctive norms in college drinking: A meta-analytic integration. *Journal of Studies in Alcohol, 64,* 331-341.

Buddin. R. (1991). *Enlistment effects of the 2+2+4 recruiting experiment* (Report R-4097-A). Santa Monica, CA: RAND Corporation.

Buddin, R., and Roan, C. (1994). *Assessment of combined active/reserve recruiting programs* (MR-504-A, 1994). Santa Monica, CA: RAND Corporation.

Campbell, D.T., and Stanley, J.C. (1966). *Experimental and quasi-experimental designs for research.* Chicago: Rand McNally.

Carroll, V. (1987). *DoD advertising mix test: Comparison of joint-service with service-specific strategies and levels of funding.* Washington, DC: Office of the Assistant Secretary of Defense (Force Management and Personnel).

Chonko, L.B., Madden, C.S., Tanner, J.F., and Davis, R. (1991). *Analysis of Army recruiter selling techniques* (Research Report 1589). Alexandria, VA: U.S. Army Research Institute for the Behavioral and Social Sciences.

Cialdini, R. (2003). Crafting normative messages to protect the environment. *Current Directions in Psychological Science, 12,* 105-109.

Colley, R.H. (1961). *Defining advertising goals for measured advertising results.* New York: Association of National Advertisers.

Colley, R.H. (1963). The power of an objective. In J.S. Wright and D.S. Warner (Eds.), *Speaking of advertising* (pp. 127-137). New York: McGraw-Hill.

Daula. T., and Smith, D. (1985). Estimating supply models for the US Army. In R. Ehrenberg (Ed.), *Research in labor economics* (Vol. 7, pp. 261-309). Greenwich, CT: JAI Press.

Davis, J.J. (1997). Concept and benefit testing. In J.J. Davis, *Advertising research: Theory and practice*. Upper Saddle River, NJ: Prentice-Hall.

Dertouzos, J.N. (1985). *Recruiter incentives and enlistment supply* (Report No. R-3065-MIL). Santa Monica, CA: RAND Corporation.

Dertouzos, J.N. (1989). *The effects of military advertising: Evidence from the Advertising Mix Test* (Report No. N-2907-FMP). Santa Monica, CA: RAND Corporation.

Dertouzos, J.N., and Garber, S. (2003). *Is military advertising effective? An estimation methodology and applications to recruiting in the 1980s and 1990s* (p. 7). Santa Monica, CA: RAND Corporation, National Defense Research Institute.

Eighmey, J. (1997). Profiling user responses to commercial web sites. *Journal of Advertising Research, May-June*, 59-66.

Federal Advisory Committee on Gender-Integrated Training and Related Issues. (1997). *Report of the Federal Advisory Committee on gender-integrated training and related issues to the Secretary of Defense*. Washington, DC: Author.

Fernandez, R. (1982). *Enlistment effects and policy implications of the Educational Test Assistance Program* (Report No. R-2935-MRAL). Santa Monica, CA: RAND Corporation.

Fishbein, M., and Ajzen, I. (1975). *Belief, attitude, intention, and behavior: An introduction to theory and research*. Reading, MA: Addison-Wesley.

Fishbein, M., Triandis, H.C., Kanfer, F.H., Becker, M.H., Middlestadt, S.E., and Eichler, A. (2001). Factors influencing behavior and behavior change. In A. Baum, T.R. Revenson, and J.E. Singer (Eds.), *Handbook of health psychology* (pp. 3-17). Mahwah, NJ: Lawrence Erlbaum Associates.

Gerrard, M., Gibbons, F.X., Reis-Bergan, M., Trudeau, L., Vande-Lune, L.S., and Buunk, B. (2002). Inhibitory effects of drinker and nondrinker prototypes on adolescent alcohol consumption. *Health Psychology, 21*, 601-609.

Goldberg, L. (1979). *Recruiters, advertising, and Navy enlistments*. Alexandria, VA: Center for Naval Analysis.

Goldberg, L, and Kimko, D. (2003). *An Army enlistment early warning system* (IDA Paper P-3783). Alexandria, VA: Institute for Defense Analyses.

Greene, W.H. (2002). *Econometric analysis*. Upper Saddle River, NJ: Prentice-Hall.

Gutman, J. (1982). A means-end chain model based on consumer characterization processes. *Journal of Marketing, 46*, 60-72.

Hastie, R., and Dawes, R. (2001). *Rational choice in an uncertain world*. Thousand Oaks, CA: Sage.

Hogan, P.F., and Smith, A.D. (1994). The accession quality cost-performance tradeoff model. In B.F. Green and A.S. Mavor (Eds.), *Modeling cost and performance for military enlistment* (pp. 105-128), National Research Council of the National Academy of Sciences. Washington, DC: National Academy Press.

Hogan. P.F., Dali, T., Mackin, P., and Mackie, C. (1996). *An econometric analysis of Navy television advertising effectiveness*. Falls Church, VA: Systems Analytic Group.

Hogan, P.F., Mehay, S., and Hughes, J. (1998, June). *Enlistment supply at the local market level*. Paper presented at the Western Economic Association annual meeting, Lake Tahoe, NV.

Hornik, R. (1997). Public health education and communication as policy instruments for bringing about changes in behavior. In M.E. Goldberg, M. Fishbein, and S.E. Middlestadt (Eds.), *Social marketing: Theoretical and practical perspectives* (pp. 45-58). Mahwah, NJ: Erlbaum.

Howard, J.A, and Sheth, J.N. (1969). *The theory of buyer behavior*. New York: John Wiley and Sons.

Hull, G.L., and Benedict, M.E. (1988). *The evaluability assessment of the recruiter training program* (Research Report 1479). Alexandria, VA: U.S. Army Research Institute for the Behavioral and Social Sciences.

Hull, G.L., Kleinman, K., Allen, G., and Benedict, M.E. (1988). *Evaluation of the U.S. Army Recruiting Command recruiter training program* (Research Report 1503). Alexandria, VA: U.S. Army Research Institute for the Behavioral and Social Sciences.

Johnson, M. (1988). Comparability and hierarchical processing in multi-alternative choice. *Journal of Consumer Research, 15,* 303-314.

Johnston, L.D., O'Malley, P.M., and Bachman, J.G. (2003a). *Monitoring the Future national survey results on drug use, 1975-2002. Volume I: Secondary school students* (NIH Publication No. 03-5375). Bethesda, MD: National Institute on Drug Abuse.

Johnston, L.D., O'Malley, P.M., and Bachman, J.G. (2003b). *Monitoring the Future national survey results on drug use, 1975-2002. Volume II: College students and adults ages 19-40* (NIH Publication No. 03-5376). Bethesda, MD: National Institute on Drug Abuse.

Johnston, L.D., O'Malley, P.M., and Bachman, J.G. (2003c). *Monitoring the Future national survey results on adolescent drug use: Overview of key findings, 2002* (NIH Publication No. 03-5374). Bethesda, MD: National Institute on Drug Abuse.

Kearl, E., Horne, D., and Gilroy, C. (1990). Army recruiting in a changing environment. *Contemporary Policy Issues, 8*(4), 68-78.

Kelly, G.A. (1955). *The psychology of personal constructs.* New York: Norton.

Kilburn, M.R., and Asch, B.J. (Eds.). (2003). *Recruiting youth in the college market: Current practices and future policy options.* Santa Monica, CA: RAND Corporation.

Kirkpatrick, D.L. (1976). Evaluation of training. In R.L. Craig (Ed.), *Training and development handbook: A guide to human resource development* (pp. 18.1–18.27). New York: McGraw-Hill.

Lehnus, J., Srokowski, S., and Daniels, G. (2000). *Importance/attribution of job traits.* Arlington, VA: Defense Manpower Data Center Market Research Program.

Mantrala, M.K. (2002). Allocating marketing resources. In B.A. Weitz and R. Wensley (Eds.), *Handbook of marketing.* Thousand Oaks, CA: Sage Publications.

McCloy, R.A., Hogan, P.F., Diaz, T., Medsker, G.J., Simonson, B.E., and Collins, M. (2001). *Cost effectiveness of Armed Services Vocational Aptitude Battery (ASVAB) use in recruiter selection* (FR-01-38). Alexandria, VA: Human Resources Research Organization.

McGuire, W.J. (1985). Attitudes and attitude change. In G. Lindzey and E. Aronson (Eds.), *Handbook of social psychology* (pp. 233-346). New York: Random House.

McQuarrie, E.F., and Mick, D.G. (1999). Verbal rhetoric in advertising: Test-interpretive, experimental, and reader-response analysis. *Journal of Consumer Research, 26,* 37-54.

Mortimer, J.T., and Finch, M.D. (1996). *Adolescents, work, and family: An intergenerational developmental analysis.* Thousand Oaks, CA: Sage.

Murray, M.P., and McDonald, L.L. (1999). *Recent recruiting trends and their implications for models of enlistment supply* (Report No. MR-847-OSD/A). Santa Monica, CA: RAND Corporation.

National Research Council. (2000). *Letter report on the Youth Attitude Tracking Study (YATS).* Committee on the Youth Population and Military Recruitment. P.R. Sackett and A.S. Mavor (Eds.). Washington, DC: National Academy Press.

National Research Council. (2002). *Letter report from the Committee on Youth Population and Military Recruitment.* Committee on the Youth Population and Military Recruitment. P.R. Sackett and A.S. Mavor (Eds.). Washington, DC: The National Academies Press.

National Research Council. (2003). *Attitudes, aptitudes, and aspirations of American youth: Implications for military recruitment.* Committee on the Youth Population and Military Recruitment. P.R. Sackett and A.S. Mavor (Eds.). Washington, DC: The National Academies Press.

Nelson, G. (1986) The supply and quality of first-term enlistees under the all-volunteer force. In W. Bowman, R. Little, and G.T. Sicilia (Eds.), *The all-volunteer force after a decade* (pp. 23-51). Washington, DC: Pergamon-Brassey's.

Office of the Assistant Secretary of Defense (Force Management Policy). (March 2000). *A new focus for military advertising and market research.* Washington, DC: Author.

Orvis, B.R., and McDonald, L. (forthcoming). *The College First/GED Plus national recruiting experiment: Results through year three.* Santa Monica, CA: RAND Corporation.

Orvis, B.R., Gahart, M.T., and Ludwig, A.K. (1992). *Validity and usefulness of enlistment intention information* (Report R-3775-FMP, p. 17). Santa Monica, CA: RAND Corporation.

Osgood, C.E., Suci, G.J., and Tannenbaum, P.H. (1975). *The measurement of meaning.* Urbana, IL: University of Illinois Press.

Otto, L.B. (2000). Youth perspectives on parental career influence. *Journal of Career Development, 27*(2), 111-118.

Overholser, C.E., and Kline, J.M. (1975). Advertising strategy from consumer research. In D.A. Aaker (Ed.), *Advertising management: Practical perspectives.* Englewood Cliffs, NJ: Prentice Hall.

Palmgreen, P., Lorch, E.P., Donohew, L., Harrington, N.G., D'Silva, M., and Helm, D. (1995). Reaching at risk populations in a mass media drug abuse prevention campaign: Sensation seeking as a targeting variable. *Drugs and Society, 8,* 29-45.

Penney, L.M., Horgen, K.E., and Borman, W.C. (2000a). *An annotated bibliography of recruiting research conducted by the U.S. Army Research Institute for the Behavioral and Social Sciences* (Technical Report 1100). Alexandria, VA: U.S. Army Research Institute for the Behavioral and Social Sciences.

Penney, L.M., Sutton, M.J., and Borman, W.C. (2000b). *An annotated bibliography of recruiting research conducted in the U.S. Navy, Marine Corps, and Air Force, and in foreign services* (Technical Report 358). Tampa FL: Personnel Decisions Research Institutes, Inc.

Polich, M., Dertouzos, J., and Press, J. (1986). *The enlistment bonus experiment* (Report No. R-3353-FMP). Santa Monica, CA: RAND Corporation.

Ramond, C. (1974). *The art of using science in marketing.* New York: Harper and Row.

Rosenstock, I.M., Strecher, V.J., and Becker, M.H. (1994). The health belief model and HIV risk behavior change. In R.J. DiClemente and J.L. Peterson (Eds.), *Preventing AIDS: Theories and methods of behavioral interventions* (pp. 5-24). New York: Plenum Press.

Rutter, M. (1980). *Changing youth in a changing society.* Cambridge, MA: Harvard University Press.

Sackett, P.R., and Mullen, E. (1993). Beyond formal experimental design: Towards an expanded view of the training evaluation process. *Personnel Psychology, 46,* 613-627.

Salas, E., Milham, L.M., and Bowers, C.A. (2003). Training evaluation in the military: Misconceptions, opportunities, and challenges. *Military Psychology, 15,* 3-16.

Sattar, K.A., Strackbein, M.E., Hoskins, J.A., George, B.J., Lancaster, A.R., and Marsh, S.M. (April, 2002). *Youth attitudes toward the Military: Poll two* (DMDC Report 2002-028). Arlington, VA: Defense Manpower Data Center.

Schwartz, B., Ward, A., Monterosso, J., Lyubomirsky, S, White, K., and Lehman, D. (2002). Maximizing versus satisficing: Happiness is a matter of choice. *Journal of Personality and Social Psychology, 83,* 1178-1197.

Sellman, W.S. (1999). *Military recruiting: The ethics of science in a practical world.* Invited address to the Division of Military Psychology, 107th annual convention of the American Psychological Association, Boston, MA.

Sellman, W.S. (2001). U.S. Military recruiting initiatives. Keynote address to the International Conference on Military Recruitment and Retention in the 21st Century. Sponsored by the Belgian Defence Staff, Royal Netherlands Army, and U.S. Office of Naval Research, The Hague, The Netherlands.

Sewell, W.H., and Hauser, R.M. (1972). Causes and consequences of higher education: Models of the status attainment process. *American Journal of Agricultural Economics, 54,* 851-861.

Shadish, W.R., Cook, T.D., and Campbell, D.T. (2002). *Experimental and quasi-experimental designs for generalized causal inference.* Boston: Houghton Mifflin.

Smith. D., Hogan, P., Chin, C., Goldberg, L., and Goldberg, B. (1990). *Army College Fund effectiveness study final report* (Office of Deputy Chief of Staff for Personnel, U.S. Army Contract No. DAKF15-87-D-0144 [SubH189-09]). Arlington, VA: Systems Research and Applications Corporation.

Spiggle, S. (1994). Analysis and interpretation of qualitative data in consumer research. *Journal of Consumer Research, 21,* 491-503.

Steinberg, L. (2003). Is decision making the right framework for research on adolescent risk taking? In Romer, D. (Ed.), *Reducing adolescent risk* (pp. 18-24). Thousand Oaks, CA: Sage.

Thornton, B., Gibbons, F., and Gerrard, M. (2002). Risk perception and prototype perception: Independent processes predicting risk behavior. *Personality and Social Psychology Bulletin, 28,* 986-999.

Todd, P., and Gigerenzer, G. (2003). Bounding rationality to the world. *Journal of Economic Psychology, 24,* 143-165.

U.S. Bureau of the Census. (2000). *Statistical abstract of the United States, 2000.* Superintendent of Documents, U.S. Government Printing Office. Washington, DC: U.S. Department of Commerce.

U.S. Bureau of the Census. (2002). *Statistical abstract of the United States,2002.* Superintendent of Documents, U.S. Government Printing Office. Washington, DC: U.S. Department of Commerce.

U.S. Department of Defense. (2002). *Population representation in the military services: Fiscal year 2001.* Washington, DC: Office of the Assistant Secretary of Defense (Force Management and Personnel).

U.S. General Accounting Office. (1998). *Military recruiting: DoD could improve its recruiter selection and incentive systems.* Washington, DC: Author.

Vinchur, A.J., Schippmann, J.S., Switzer, F.S., and Roth, P.L. (1998). A meta-analytic review of predictors of job performance for salespeople. *Journal of Applied Psychology, 81,* 586-597.

Von Neumann, J., and Morgenstern, O. (1947). *Theory of games and economic behavior.* Princeton, NJ: Princeton University Press.

Warner. J. (1990). Military recruiting programs during the 1980s: Their success and policy issues. *Contemporary Policy Issues, 8*(4), 47-57.

Warner. J. (1991). *Navy recruiting incentive models final report* (Contract No. DAAL-86-D-0001, Delivery Order 2217). San Diego: Navy Personnel Research and Development Center.

Warner, J., Simon, C., and Payne, D. (2001). *Enlistment supply in the 1990s: A study of the Navy College Fund and other incentive programs* (DMDC Report No. 2000-015). Arlington, VA: Defense Manpower Data Center.

Warner, J., Simon, C., and Payne, D. (2002). *Propensity, application, and enlistment: Evidence from the Youth Attitude Tracking Survey*. Clemson, SC: Clemson University Department of Economics.

Wilson, M.J., Greenlees, J.B., Hagerty, T., Helba, C.V., Hintze, D.W., and Lehnus, J.D. (2000). *Youth Attitude Tracking Study 1999: Propensity and advertising report* (C-DASW01-96-C-0041). Arlington, VA: Defense Manpower Data Center.

Zuckerman, M. (1979). *Sensation seeking: Beyond the optimal level of arousal*. Hillsdale, NJ: Erlbaum.

Appendix

Biographical Sketches

Paul R. Sackett (*Chair*) is professor in the Department of Psychology at the University of Minnesota, Twin Cities. His research interests revolve around legal, psychometric, and policy aspects of psychological testing, assessment, and personnel decision making in workplace settings. He has served as the editor of *Personnel Psychology*, as president of the Society for Industrial and Organizational Psychology, as cochair of the Joint Committee on the Standards for Educational and Psychological Testing, as a member of the National Research Council's Board on Testing and Assessment, and as chair of the American Psychological Association's Board of Scientific Affairs. He has a Ph.D. in industrial and organizational psychology from Ohio State University.

David J. Armor is professor of public policy in the School of Public Policy at George Mason University, where he is director of the Ph.D. program. He also teaches statistics and social policy and conducts research in education, military manpower, and family policy. He began his research in military manpower while at the Rand Corporation. Between 1986 and 1989 he served as principal deputy and acting assistant secretary for Force Management and Personnel in the U.S. Department of Defense. He was a member of the National Research Council's Committee on Military Enlistment Standards. He has a Ph.D. in sociology from Harvard University.

Jerald G. Bachman is program director and distinguished senior research scientist in the Survey Research Center of the Institute for Social Research at the University of Michigan, Ann Arbor. His scientific publications focus on youth and social issues. His current research interests include drug use and attitudes about drugs; youth views about military service; other values, attitudes, and behaviors of youth; and public opinion as related to a number of other social issues. He is a principal investigator on the Monitoring the Future project and the principal investigator on the Youth Attitudes About Military Service project. He has a Ph.D. in psychology from the University of Pennsylvania.

Marilyn Dabady (*Senior Research Associate*) is study director for the National Research Council's Panel on Methods for Assessing Discrimination in the Committee on National Statistics. Her background is in social psychology, organizational behavior, and human resource management. Currently, her main areas of interest are interpersonal and intergroup relations; prejudice, stereotyping, and discrimination; and diversity management. She has a bachelor's degree in psychology from the University at Albany, State University of New York, and M.S. and Ph.D. degrees in psychology from Yale University.

John Eighmey holds the Raymond O. Mithun Land Grant Chair in Advertising in the School of Journalism and Mass Communication at the University of Minnesota. He is an authority on advertising and marketing, consumer research, and the management of strategic communication programs. He has held senior management positions at Young & Rubicam, a worldwide advertising agency based in New York City, and at the Federal Trade Commission in Washington, DC. He has a Ph.D. in marketing from the University of Iowa.

Martin Fishbein is the Harold C. Coles Distinguished Professor of Communications in the Annenberg School at the University of Pennsylvania. His areas of expertise include attitude theory and measurement, communication and persuasion, behavioral prediction and change, and intervention development, implementation, and evaluation. He also has carried out studies of the relations among beliefs, attitudes, intentions, and behaviors in field and laboratory settings. He has a Ph.D. in psychology from the University of California at Los Angeles.

Carolyn Sue Hofstrand is the director of counseling and guidance at Taylor High School in Volusia County, Florida. She is a nationally certified school counselor with experience at elementary, middle, high, and postsecondary schools. She was named National Secondary School Counselor of the Year 2000 by the American School Counselor Association and has been involved in leadership positions in counseling organizations at the state, local, and national levels. She has a master's degree in education, counseling, and guidance from North Dakota State University, where she serves on the board of visitors for the College of Education and Human Resources.

Paul F. Hogan is vice president and senior economist at The Lewin Group in Fairfax, Virginia. He has more than 20 years of experience in applying microenonomics, statistics, and operations research methods to problems in labor economics, including labor supply and demand, efficient staffing methods, and performance and cost measurement. He served as the senior analyst on the President's Military Manpower Task Force and as director of Manpower Planning and Analysis in the Office of the Secretary of Defense, the office charged with staffing methods and criteria used by military departments to determine demands for personnel. He was awarded the Secretary of Defense Distinguished Civilian Service medal in 1982 and 1985, and the Navy Superior Civilian Service medal in 1980. His doctoral studies include economics, econometrics, and finance at the University of Rochester and his undergraduate degree is in economics from the University of Virginia.

James Jaccard is a distinguished professor of psychology at the State University of New York, Albany. He areas of expertise include experimental design and statistical analyses. He has written several books on statistical analysis. His work over the past 20 years has included research studies on the effects parental influence on youth decision making and on the formation of youth attitudes as a function of information input. He has a Ph.D. in social psychology with a minor in quantitative methods and cognitive psychology from the University of Illinois.

Carolyn Maddy-Bernstein is associate to the executive vice president at the University of Arizona where she directs the University's diversity initiatives. Her expertise includes research and service in career and technical education, guidance and counseling, and educating students who are members of underrepresented groups. From 1988 to 1999, she served as director of the Office of Student Services for the National Center for Research in Vocational Education at the University of Illinois, Urbana-Champaign. As a Louisiana State University faculty member, she taught and worked on research and service projects funded by the Governor's Office of Workforce Development and the Department of Education. She has been a public school teacher, counselor, and administrator at the secondary and postsecondary levels. She has a Ph.D. in education from Virginia Polytechnic Institute and State University.

Anne S. Mavor (*Study Director*) is the staff director for the Committee on Human Factors and the Committee on the Youth Population and Military Recruitment. Her previous National Research Council work has included studies on occupational analysis and the enhancement of human performance, modeling human behavior and command decision making, human factors in air traffic control automation, human factors considerations in tactical display for soldiers, scientific and technological challenges of virtual reality, emerging needs and opportunities for human factors research, and modeling cost and performance for purposes of military enlistment. For the past 25 years, her work has concentrated on human factors, cognitive psychology, and information system design. She has an M.S. in experimental psychology from Purdue University.

Carol A. Mutter is a retired lieutenant general of the United States Marine Corps. Her experience has been in research, development, and acquisition, as well as financial management, logistics, personnel administration, and equal opportunity. In her most recent Marine Corps assignment, she was the senior Marine Corps personnel management executive, making policy for and managing the careers and quality of life of all Marines and civilians working for the Marine Corps. She is the chair of the Defense Advisory Committee on Women in the Services, as well as currently serving on the National Advisory Council of the Alliance for National Defense, the Advisory Board for the Indiana Council on World Affairs, and is the National President of the Women Marines Association, as well as a senior fellow at the Joint Forces Staff College. She has an M.A. in national security and strategic studies from the Naval War College in Newport, RI.

Luther B. Otto is William Neal Reynolds Distinguished Professor of Sociology emeritus at North Carolina State University, Raleigh. His research focuses on youth and careers. He directed the Career Development Study, a detailed study of the early career histories of 7,000 young men and women from the time they were juniors in high school through age 30. He has published numerous articles, chapters, and books on youth and careers. He served two four-year terms on the Basic Social-Cultural Research Review Committee of the National Institutes of Health. He has been active in professional associations, has served in a number of editorial capacities, and frequently consults with federal and state agencies and private foundations on issues related to youth, education, and work. He has a Ph.D. in sociology from the University of Wisconsin-Madison.

William J. Strickland is vice president of the Human Resources Research Organization (HumRRO) in Alexandria, Virginia. He also directs its

Workforce Analysis and Training Systems Division. He is a retired Air Force colonel who was director of human resources research at the U.S. Air Force Armstrong Laboratory. In that position, he was responsible for all Air Force research in the areas of manpower and personnel, education and training, simulation and training devices, and logistics. Earlier in his career, he commanded an Air Force recruiting squadron, was the chief of market research for Air Force recruiting, and was the deputy director for operations for Air Force recruiting. A fellow of the American Psychological Association, he is a past president of its Division of Military Psychology. He has a Ph.D. in industrial and organizational psychology from the Ohio State University.

Nancy T. Tippins is president of the Selection Practice Group of Personnel Research Associates, Inc., in Arlington Heights, IL. She is responsible for the development and execution of the firm's strategies related to employee selection, assessment, and development. Prior to joining the firm, she spent over 20 years managing personnel research functions involved in selection methods, staffing policies and procedures, equal employment opportunity and affirmative action, outplacement and downsizing, human resource services, and surveys for GTE, Bell Atlantic, and Exxon Company, USA. She has a Ph.D. in industrial and organizational psychology from the Georgia Institute of Technology.

John T. Warner is professor of economics at Clemson University. He was a member of the 9th Quadrennial Review of Military Compensation with the Office of the Undersecretary of Defense for Personnel and Readiness. He has written extensively on the topic of economic incentives and cost analysis in military recruitment with particular emphasis on the relative cost-effectiveness of various benefit packages (pay, bonuses, educational programs, etc.) on raising enlistments. He has a Ph.D. in economics with a minor in statistics from North Carolina State University.

Index

B

C

S

T

U

W

Y